Biochemistry of Aging

Wellness and Longevity

By

Amin Elsersawi, Ph.D.

authorHOUSE®

AuthorHouse™
1663 Liberty Drive
Bloomington, IN 47403
www.authorhouse.com
Phone: 1-800-839-8640

First published by AuthorHouse 1/22/2010

ISBN: 978-1-4490-7380-0 (sc)
ISBN: 978-1-4490-7381-7 (e)

Library of Congress Control Number: 2010900951

Printed in the United States of America
Bloomington, Indiana

This book is printed on acid-free paper.

Appendix

Dedication

This book is dedicated to my loving parents who gave me life.
Always I think about you, wondering the future will take us.
You are the best companion I can imagine.

**Do not regret growing older.
It is a privilege denied to many.
~Author Unknown**

This book is not designed to provide medical advice or professional services. It is intended to be for educational and demonstrational use only. The information and thoughts provided in this book are not a substitute for professional care and should not be used for diagnosing, establishing, or treating a health problem or a disease. Your doctor is the only person you should consult if you have, or suspect a health problem with your aging.

Introduction

On July 1, 2004, 12 percent of all Americans were aged 65 and over. The Census Bureau reports that by 2050, people 65 and over will comprise an impressive 21 percent of the U.S. population. Every year since May 1963, Older Americans Month has been honored with a presidential proclamation. President Bush said, "Older Americans help others understand the past, and they teach timeless lessons of courage, endurance and love. Through their legacy of patriotism, service, and responsibility, America's seniors also unite families and communities and serve as role models for younger generations."

Aging is not a process of decline; it is a process of continuing to meet life's challenges and growing into a complete human being. Aging is a complex interaction of genetics, chemistry, biology and physiology; topics that are worthwhile to address.

This book offers a good introduction to the biology and chemistry of aging. It emphasizes on cellular aging, and covers different areas and theories which deal with mechanism of aging. If the reader has some background in biology, then this is an excellent introductionary book to the biology and chemistry of aging.

This book includes information on aging of cells, reversal effect of aging, DNA damage theory of aging, acidity, oxidative stress, radicals, insulin/IGF hormones, anabololic and catabolic hormones, testosterone and estrogen, energy through conversion between NAD+ and NADH, DNA transcription to RNA, and RNA translation to protein.

This book is the first of its kind in providing information for scientists, physicians, pharmacists, engineers, teachers, computer programmers, and anyone with a background or strong interest in biology, chemistry, and aging. The book deals with scientific causes of aging, and philosophical and sociological implication of life-extension and research on aging. I recommend it for those serious about the biology and chemistry of aging. This book has three chapters:

Biology

The contents of this chapter include information on molecular and cell biology, physiological biology with details on the membrane, mitochondrion, microbules, cytoskeletal system, protein synthesis, and nucleus. It includes discussions on the centrosome cycle, components of the cell, NAD, ATP, and ADP and also ADP and ATP exchange, the glycolysis process, Kreb cycle, amino acids and steriosomers of amino acids.

Chemistry

This chapter includes information on atomic orbitals, polarity of molecules, electronegativity, ionic and covalent bonding, isotopes and reactivity, chemical structures, isomerism, resonance, hydrogen bonding, IUPAC nomenclature, ionization, chemical reactivity, acidity and alkalinity, amino acids, names of organic compounds, functional group, molecular biology, carbohydrates, lipid, protein, and nuclear medicine and radioactivity.

Aging

This chapter includes information on the aging of cells and the reversal of aging, the DNA damage theory of aging, acidity, oxidative stress, radicals, insulin/IGF hormones, anabololic and catabolic hormones, testosterone and estrogen, energy through conversion between NAD+ and NADH, DNA transcription to RNA, and RNA translation to protein. Topics such as DNA mutation and methylation, gene expression and gene silencing, inheritance and aging are also addressed. This chapter talks and expresses public arguments about social and political issues about gerontology.

This chapter ends with the author's exclusive opinion about causes of aging and elongation of life span.

Preface

This book is about how your body works, and about the chemical reaction involved inside of it. Discussing the biology and the chemistry of your body may help you understand how aging develops, and how, when armed with knowledge and an enlightened strategy, we can safely recapture the energy and mental capability even the body of our youth.

In this book, biology will deal with the activities and characteristics of all organisms in human which fall into two major categories: reproduction and metabolism. The mechanism of reproduction is now known to be controlled by the properties of certain large molecules called nucleic acids. They transcribed the entire DNA helix at once into mRNA and also the cross selection between alleles (which control the same inherited characteristics) in both parents.

The other major activity of a living organism is metabolism. It is the physical, chemical, and physiological processes by which energy and synthesis of proteins, hormones, and enzymes are used in such activities such as reproduction (including growth), activities, and responsiveness to the environment, (which also constitutes the activities of the nervous system).

The main core of the book consists of Chapter 3. The chapter points out that the author believes that although aging is an irreversible process, there are many things we can do to keep our minds and bodies in good working order through all phases of life. This can be accomplished by understanding the biology and chemistry of the human body. The materials in this chapter can help you achieve and maintain the best health throughout the lifelong process of aging. Read chapters 1 and 2 before you skip to chapter 3.

CHAPTER   1

BIOLOGY

Introduction

Biology is the study of life. Biology is such a broad field, covering the structure, function, growth, origin and how they interact with each other and the environment that they live in. It is the branch of knowledge which treats living matters as distinct from non-living matters. Biology can be divided into groups:

- Botany - the study of plants
- Zoology - the study of animals
- Microbiology - the study of microorganism
- Biochemistry - the study of chemistry of life.

All above groups are further divided into subgroups:

- Molecular biology - the study of interactions of biological molecules
- Cellular biology - the study of the basic block of life
- Physiological biology - the study of physical and chemical interactions of cells
- Ecology-the study of the interaction of organs and the environment

The foundation of biology starts with understanding cellular biology which mainly covers seven topics as shown in the diagram below:

The cell is the fundamental of life, and all living things are composed of one or more cells. Cell theory provided a new perspective on the fundamental basis of life. Cells are subjected to mutations and development based on five principles; cell theory, evolution, gene theory, homeostasis, and energy.

The study of molecular biology also includes microscopic anatomy (histology) which can be examined by the aid of microscopes, dissection and inspection of dead bodies (cadavers). Human anatomy, biochemistry, and physiology are subjects of important discussion within biology. Anatomy involves several subjects such as: embryology, osteology, syndesmology, myology, angiology, the arteries, the veins, the lymphatic system, neurology, the organs of the senses and the common integument, splanchnology, surface anatomy and surface marking.

Medical doctors, especially surgeons and histopathologists and radiologists should have a thorough working knowledge of anatomy.

1-1 Cell biology

Cell age is dependent on the number of divisions a cell has undergone, not its chronological age. Cells typically go into senescence after about 50 divisions (Hayflick's limit), and are eventually removed by apoptosis. But most cells do not reach Hayflick's limit.

The cell is the smallest structural and functional unit of all living organisms. It is often called the building block of life. Some organisms are unicellular such as ameba and bacteria, and some are multicellular such as humans. Humans have about 100 thousand billions of cells with a diameter of about 10 micrometers and of a weight of 1 nanogram, Figure (1).

Figure (1): One cell with cytoplasm and nucleus

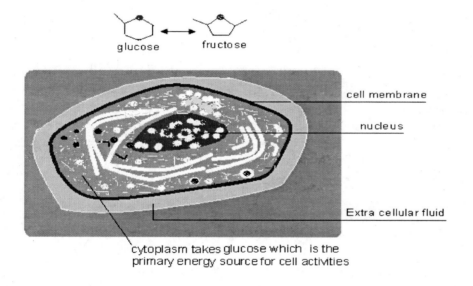

cytoplasm takes glucose which is the primary energy source for cell activities

The Cytoplasm needs energy that is taken from glucose. Glucose and fructose have a similar number of carbon, hydrogen, and oxygen. Both have the formula $C_6H_{12}O_6$. The difference is that glucose is hexagonal and has only one double bonded oxygen, and fructose is pentagonal and has two double bond of oxygen. Both are isomers and interchangeable under the process of photosynthesis. In addition to the carbohydrates, lipids, and proteins, the cell needs something called nucleic acid. Carbohydrates have a unique formula which is the ratio between carbon, hydrogen, and oxygen which is about 1:2:1. Lipids, protein, and nucleic acids have different ratios. We discussed carbohydrates, lipids, and proteins in full details in chapter 1.

Nucleic acid is used by the cell to store and use hereditary information. Nucleic acids known as nucleotides consist of three main components: a nitrogenous group, a phosphate group, and a sugar:

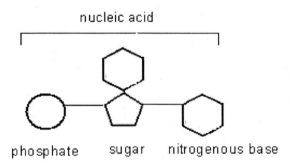

There are two types of nucleotides distinguished by their sugar: Deoxyribonucleic acid (DNA), which has one less oxygen than the other type ribonucleic acid (RNA). The function of DNA molecules is to store genetic information for the cell. RNA molecules carry genetic messages from the DNA in the nucleus to the cytoplasm for use in protein synthesis and other processes. We shall discuss nucleic acids in detail in the coming sections.

1-1-1 Membranes

There are several membranes associated with the cell. There are, for example, membranes to protect the cell from outside environment, membranes to insulate nerve fibers (which are called myelin), plasma membranes, and mitochondrial membranes. All membranes contain lipids and proteins with different ratios. Plasma has a lipid over protein ratio of 43.53/44.49, and the mitochondrial membrane of 24/76. As mentioned before in chapter 1 lipid has a hydrophilic head (loves water), and a tail which is hydrophobic (hated water), Figure (2).

Figure (2): Cross section of a cell

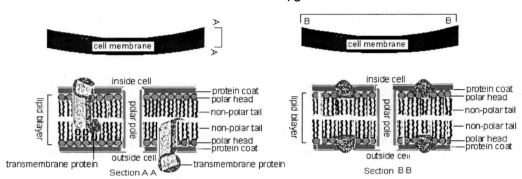

Section A A   Section B B

Transmebrane proteins may be embedded in the lipid bilayers, on the edges of the membrane inside the cytoplasm or on the exterior face of the membrane.

In addition to the normal lipid (saturated lipid) there is an important type of lipid in the membrane is called the phospholipids in addition to the normal lipid (saturated lipid). These phospholipids have a polar head group and two hydrocarbon tails; one tail is saturated and the second is unsaturated because two carbons are double bonded as shown in Figure (3).

Figure (3): Phospholipids of one chain saturated and the other isdouble bonded-carbon, see chapter 2.

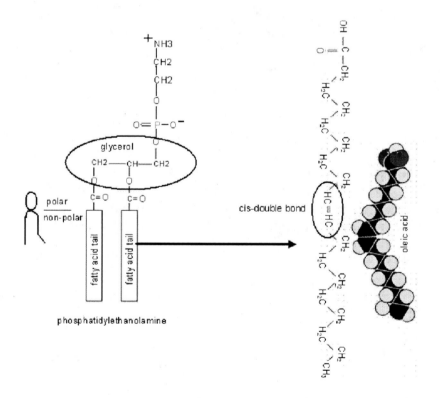

In Figure (3), one can see a kink in the double bond chain which prevents the bilayer tight packing and also makes the bilayer difficult to freeze. The effect of the kink and the floating of the polar head of the phospholipids make it fluid characteristics. It means that at low temperature, the membrane is in a gel state and slightly compressed. At higher body temperature the inside of the membrane is in a fluid state which allows the lipid molecules to move around, rotate, and exchange places. This allows movement of other components of the membrane, http://www.cytochemistry.net/Cell-biology/membrane_intro.htm.

Surrounding each of our cells is a membrane called the plasma membrane. The plasma membrane is a continuous double-layer of phospholipids, interweaved with cholesterol and proteins. Membranes have also cholesterol of different type of lipid. Different ratios between cholesterol and phospholipid molecules are in different membranes. The ratio in plasma membranes is 1:1. In cell membranes cholesterol can make up nearly half of the cell membrane. In bacteria the ration is 0:1 (bacteria have no cholesterol). Cholesterol is also present in membranes of organelles inside the cells, although it usually makes up a smaller proportion of the membrane. For example, the mitochondrion, the so-called "power-house" of the cell, contains only three percent cholesterol by mass. The endoplasmic reticulum (ER), Figure (4), which is involved in making and modifying proteins, is six percent cholesterol by mass. Since cholesterol molecule is smaller and weighs less than other molecules in the cell membrane, it makes up a lesser proportion of the cell membranes mass (roughly 20 percent). http://www.cholesterol-and-health.com/Cholesterol-Cell-Membrane.html.

Cholesterol consists of four hydrocarbon rings, which are strongly hydrophobic, and a hydroxyl (OH) group which is attached to one end of the cholesterol chain. Hydroxyl is known to be hydrophilic, meaning that cholesterol is amphiphatic. Cells produce cholesterol or draw it from extracellular sources with lipoprotein when it is needed, as cholesterol is continuously lost to the outside processes. The endoplasmic reticulum synthesizes the cholesterol which then binds to cellular proteins called caveolins in a 1:1 ratio. Both are involved in transport of de novo-synthesized cholesterol from the ER to the plasma membrane. Cellular homeostasis of cholesterol involves in the regulation of its amount and its movement between membranes. Cholesterol is therefore an important compound that affects the properties and functions of membrane proteins such as enzymes, receptors, or ionizing molecules for transportations and communications. Thus cholesterol helps maintain the integrity of membranes, and plays a role in facilitating cell signaling and communications.

Figure (4): cell and endoplastic reticulum

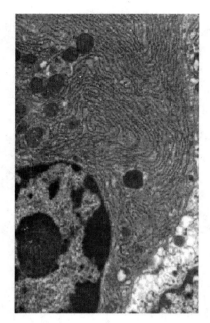

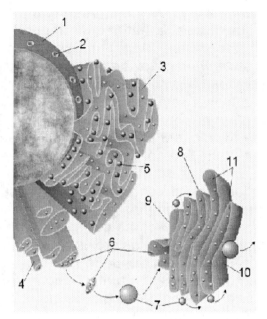

Micrograph of rough endoplasmic reticulum network around the nucleus (shown in lower right-hand side of the picture). Dark small circles in the network are mitochondria

1 Nucleus **2** Nuclear pore **3** Rough endoplasmic reticulum (RER)
4 Smooth endoplasmic reticulum (SER) **5** Ribosome on the rough ER
6 Proteins that are transported **7** Transport vesicle **8** Golgi apparatus
9 Cis face of the Golgi apparatus **10** Trans face of the Golgi apparatus
11 Cistemae of the Golgi apparatus

http://en.wikipedia.org/wiki/Endoplasmic_reticulum

Recent research added a new dimension to the mechanism of cholesterol in interaction with mycobacterium such as tuberculosis and leprosy. Mycobacterium possesses cholesterol receptor, say C_m that tries to enter the macrophage of a cell through the cell membrane. A human cell receptor, say C_h, regulates the transcriptional expression of a gene that codes for Tryptophan-Asperate containing coat (TACO) protein which is responsible for the survival of mycobacterium within cells, Figure (5). The two receptors C_m and C_h with cholesterol-rich membrane domains helps to create a junction between mycobacterium and macrophage. This results in allowing mycobacterium to enter and survive within macrophages so they will not deplete into the cell. Gatfield and Pieters reported that TACO protein, associated with the membrane cholesterol and cell depleted cholesterol, were unable to internalize mycobacteria. The molecular mechanism through which cholesterol initiates this process is still poorly understood. However, agents can be designed to target the cholesterol-mediated entry of M. tuberculosis may be developed as tuberculosis therapeutics,
http://cat.inist.fr/?aModele=afficheN&cpsidt=15473877

Figure (5): TACO protein and cholesterol molecules encircle bacterium

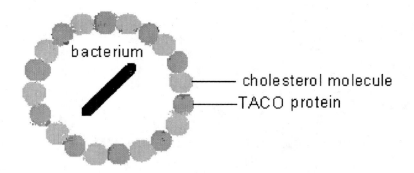

bacterium

cholesterol molecule

TACO protein

1-1-2 Mitochondria

Mitochondria are central components of our cells that generate the majority of our energy from nutrients. But their weak side is that, through their normal activity, they generate unstable chemicals that harm both the mitochondrion itself and other components of the cell. This resulting damage is thought to play a important role in aging.

Mitochondria are called the powerhouses of the cell. They are membrane enclosed organelles found in most eukaryotic cells (eukaryotic cells are in all animals except bacteria and cyanobacteria). They absorb the nutrients, break them down and then release energy called ATP, for the cell. The process of releasing energy is known as cellular respiration, which is associated with chemical reactions needed for the growth of the cell. Mitachondria are very small organelles. Some cells has one mitochondrion and others have several thousand mitochondria. Cells for transmitting nerve impulses have fewer mitochondria than muscle cells that need high energy. The numbers of mitochondria are adjusted depending on the energy needed by the cell. Mitochondria can move, grow, and combine with other mitochondria inside the cell. Figure (6) is of a mitochondrion with the outer and inner membranes. The purpose of Cristae is to increase the surface area inside the cell for more chemical reaction, when required. The matrix inside the mitochondria is fluid. Inside the matrix, oxygen is combined with protein to digest the food molecules, and the water inside the matrix, takes the digested food to feed the cell. Mitochondria are the only places in the cell where oxygen is used to digest the food molecules. Calcium may also be controlled by mitochondria.

Mitochondria have their own independent genome, the material of which is known as mitochondrial DNA (mtDNA) which we'll talk about it.

Figure (6): A human mitochondrion with its DNA, cristae, matrix, and ribosome

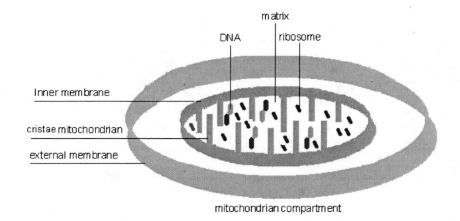

One cell may have several thousand mitochondria, Figure (7).

Figure (7): A micrograph of one human cell with mitochondria around the nucleus

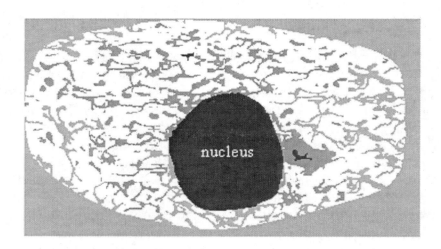

Another diagram, Figure (8), shows most of the cell's components, and a sketch of the nucleus.

Figure (8): Components of a human cell

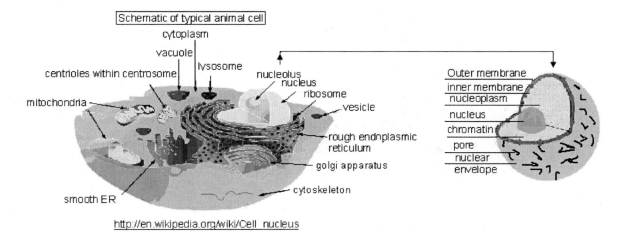

Schematic of typical animal cell

cytoplasm

vacuole

lysosome

centrioles within centrosome

nucleolus
nucleus
ribosome

mitochondria

vesicle

rough endoplasmic reticulum

golgi apparatus

cytoskeleton

smooth ER

Outer membrane
inner membrane
nucleoplasm
nucleus
chromatin
pore
nuclear
envelope

http://en.wikipedia.org/wiki/Cell_nucleus

We shall discuss the components of the cell that are shown in Figure (8) above.

1-1-3 Nucleus

The cell nucleus is a huge storeroom of biochemical information. It is the brain of the cell and the core for direction and coordination of the cell's reproductive and metabolic activities. The nucleus is the largest organelle of a cell. Cells with a nucleus are called eukaryotic cells because eukaryotic means, "possessing a true nucleus immersed inside the cell". Plants and animals have nuclei, where organisms such as bacteria do not. The nucleus is about two to five micrometers long in diameter. It is surrounded by two membranes that form the nuclear envelope, see Figure (9). Most of the cell's genes are located in the nucleus. The nucleus contains DNA, RNA, a nucleoplasm, and a nucleolus. The two main parts of the nucleus are the nucleolus and the nucleoplasm. The nucleolus is usually visible as a dark, round spot in the nucleus. It helps in the formation of ribosomes. The nucleolus is made up of proteins and RNA. It is the site of ribosomal ribonucleic acid (rRNA) synthesis.

Figure (9): Components of a nucleus

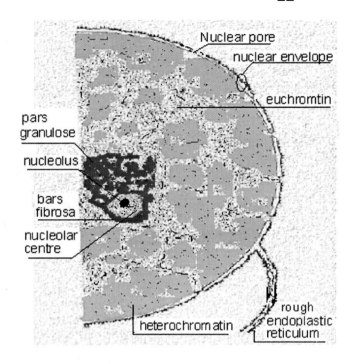

1-1-4 Cytoplasm

There are two parts of the cytoplasm of a cell; the upper part and the lower part. The upper part contains organelles such as the nucleus, the mitochondria, centrosomes, and ribosome. The lower part is the cytosol which is a complex mixture of the cytoskeleton, dissolved molecules, filaments, and water that fills much of the volume of a cell. The cytoskeleton is a network of protein filaments that are responsible for the movement and support of the cell. It also and give the cell its shape and outline. The cytoplasm has many different molecules dissolved in solutions such as fatty acids, sugars, and amino acids that are used to keep the cell working. Waste products are also dissolved in the cytoplasm before they are taken in by vacuoles and sent out of the cell. Another different cytoplasm is called the nucleoplasm, which is of different composition and only found inside the nucleus. The nucleoplasm holds the cell's chromatin and nucleolus. It is not always present in the nucleus. When the cell divides, the nuclear membrane dissolves and the nucleoplasm is released. After the cell nucleus has reformed, the nucleoplasm fills the space again.

The Cytoplasm is made up of proteins, amino acids, vitamins, nucleic acids, ions, carbohydrates, sugars, and fatty acids. All of the functions for cell expansion, growth and replication are carried out in the cytoplasm.

1-1-5 Lysosomes

Lysosomes are the cells' garbage disposal system. They are membrane-bound vesicles that contain hydrolytic enzymes that digest particles or cells such as bacteria taken into the cell by phagocytes. Hydrolytic enzymes are formed in the endoplasmic reticulum by the end result of the interaction of proteins, nucleotic acids, lipids and carbohydrates in the cell. After the bacterium is engulfed in a vacuole (a small cavity in the membrane or the cytoplasm), vesicles containing lysosomal enzymes combine with it. The destruction of bacteria releases hydrogen and reduces the pH of the enzyme of the lysosome. The effect of the reduction in the pH activates the enzymes. The lysosomes are then split into more secondary lysosomes and this results in the destruction and recycling of the bacteria. Lysosomes also degrade worn out mitochondria. The process of digestion is accomplished in three stages: bacteria or foreign bodies that enter the membrane will be encircled by the membrane itself, then the encircled membrane will be pinched off from the cells outer membrane to produce an endosome,.Then the lysosomes will fuse with the endosome to degrade the contents.

Figure (10) shows the membranes, lysosomes, mitochondria, components of endoplasmic reticulum, and Golgi apparatus.

There are other garbage disposal systems called "peroxisomes" which are similar to the lysosomes, but they are much more dynamic and can replicate by enlarging and then dividing. Peroxisomes have enzymes that rid the cell of toxic peroxides such as Hydrogen peroxide (H_2O_2).

Figure (10): Lysosomes and other interacted components

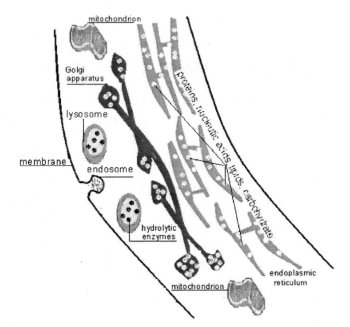

1-1-6 Ribosomes

Ribosomes are the protein synthesizers or the protein builders of the cell they connect amino acids in long chains. Ribosomes are found in many places around the nucleus of the cell. They are floating in the cytoplasm (cytosol). Some ribosomes are found on the endoplasmic reticulum and they are called the rough endoplasmic reticulum, see Figure (9) above. A ribosome has two separated pieces or subunits called 60-S (large) and 40-S (small). When the cell needs to make protein, mRNA (the messenger RNA) is created in the nucleus. The mRNA is then sent into the cell and the ribosomes. When it is time to make the protein, the two subunits come together and combine with the mRNA. The subunits lock onto the mRNA and start the protein synthesis, http://www.biology4kids.com/files/cell_ribos.html.

The 60-S/ 40-S model works fine for eukaryotic cells. Prokaryotic cells have ribosomes made of 50-S and 30-S subunits. It's a small difference. Scientists have used this difference in ribosome size to develop drugs that can kill prokaryotic microorganisms that cause diseases, Figure (11).

Figure (11): Subunits of one ribosome; 60-s and 40-s ribosome subunits (two)

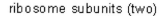

ribosome subunits (two)

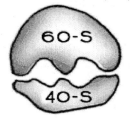

Another nucleic acid lives in the cell - tRNA (transfer RNA). The tRNA is bonded to the amino acids floating around the cell. With the mRNA offering instructions, the ribosome connects to a tRNA and pulls off one amino acid. Slowly the ribosome makes a long amino acid chain that will be part of a larger protein, Figure (12).

Figure (12): Formation of protein (translation), see chapter 3

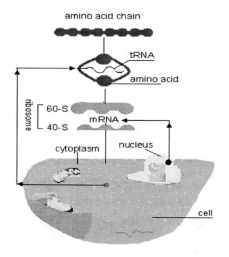

Vesicles: a bubble-like membranous structure that stores and transports cellular products, and digests metabolic wastes within the cell. It is an intracellular membranous sac that is separated from the cytosol by at least one lipid bilayer that consists of a hydrophilic head and hydrophobic tail, Figure(13).. Vesicles are involved in metabolism, transport, and enzyme storage. They can also act as being chemical reaction chambers. Vesicles move molecules between locations inside the cell. For example, proteins from the rough endoplasmic reticulum, see Figure (9), to the Golgi apparatus and then to the lysosomes. Such proteins mature in the Golgi apparatus before they move by the transport vesicles to the lysoomes, Figure (13).

Figure (13): Shape of a vesicle

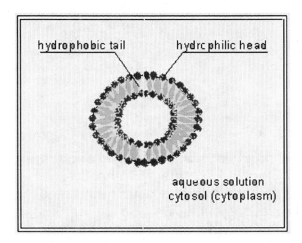

1-1-7 Rough endoplasmic reticulum

The endoplasmic reticulum has two types: the rough endoplasmic reticulum and the smooth endoplasmic reticulum (RER and SER). Both are of networks of tubules, vesicles and sacs that are interconnected. They serve specialized tasks in the cell such as protein synthesis, sequestration of calcium, storage and production of glycogen, production of steroids, and insertion of membrane proteins. Most of these jobs are the responsibility of the rough endoplasmic reticulum, which gets its name (rough) from the presence of ribosomes on its surface. The rough endoplasmic reticulum carries the ribosomes during protein synthesis. The newly synthesized proteins are sequestered in sacs, called cisternae. The system then sends the proteins via small vesicles to the Golgi apparatus, Figure (14), Or in the case of membrane proteins, it inserts them into the membrane.

Figure (14): Sequence of protein synthesis

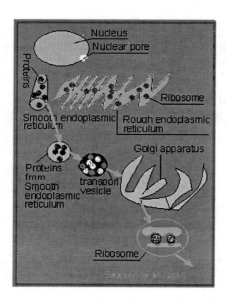

The smooth endoplasmic reticulum has no bound ribosomes like the rough endoplasmic reticulum, and therefore is not involved in protein synthesis. The smooth endoplasmic reticulum involves in cholesterol metabolism, membrane synthesis, detoxification, and storage of the cations of calcium; Ca+. The SER is the recipient of the protein that synthesized in the RER. These proteins will be exported and passed to the Golgi apparatus. The treated proteins in the Golgi apparatus are then sent back to the RER. Similarly, proteins come from lysosomes. After phosphorylation of their mannose residues (sugar $C_6H_{12}O_6$) they will go back to lysosomes. Glycosylation (production of polysaccharides that modulate the structure and function of proteins) of the glycoproteins (proteins that contain polysaccharides; glycans covalently bonded to the proteins' end chains, will also continues together with the synthesis of lipids in the SER. The membrane of the SER also contains enzymes that catalyze a series of reactions to detoxify both lipid-soluble drugs and harmful products of metabolism. Large quantities of certain compounds such as Phenobarbital cause an increase in the amount of the smooth ER, [source: ISBN:0198506732].

1-1-8 Golgi apparatus

The Golgi apparatus is also called the Golgi complex, or Golgi body. It is comprised of a series of five to eight, cup-shaped, membrane-covered sacs called cistemae. The Golgi apparatus is made up of about 60 cistemae. The Golgi is considered the dispatch centre for the transport of the cell's chemical products to other locations inside the cell, or outside the cell, Figure (15). It modifies proteins and lipids that have been built in the endoplasmic reticulum, and transports them as required. The Golgi is principally responsible for directing molecular traffic inside and outside the cell. Almost all molecules pass through the Golgi apparatus at some point in their existence. The arrangement of the transport is mediated by the vesicles. When proteins bind with their appropriate receptor on the vesicle, they are distinct in the vesicle and transported away.

Figure (15): Golgi apparatus

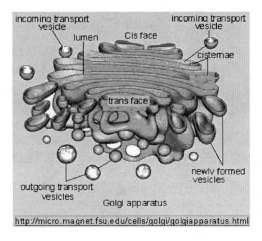

1-1-9 Cytoskeleton

Cytoskeleton has the functions of:
- intercellular transport of organelles
- establishing cell shape, and providing physical strength
- separation of chromosomes during mitosis and meiosis
- movement of the cell and other organelles from one place to another

The cytoskeleton is made up of three kinds of proteins: (The main associated proteins here are kinesin and myosin), Figure (16).
- actin filaments (microfilaments)
- intermediate filaments
- microtubules

The cytoskeleton exists in eukaryotic cells to perform a variety of shapes and to carry out coordinated and directed movements of the cells and organelles. Cytoskeletons are not found in bacteria. Since all cells are polar, Cytoskeletons are used to differentiate parts of the cell by the microtubules associated with cytoskeletons. Cytoskeletons play an important role in cell polarity because of their own intrinsic polarity.

The actin filaments are found in muscles where they contract and expand. Running, walking, swimming, and flying depend on the rapid contraction of the microfilaments, while the pumping of the heart and gut peristalsis (movement of voluntary muscle) depend on the contraction of cardiac and smooth muscle.

Figure (16): Cytoskeleton with microfilament and microtubules

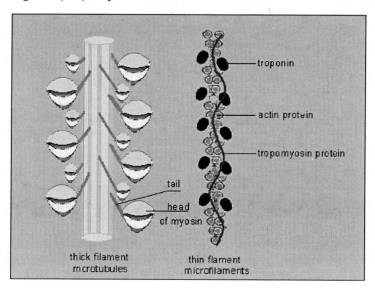

1-1-10 Vacuole

A vacuole is a membrane-bound sac that contains water and waste products. It plays roles in intercellular digestion and the release of waste products due to the digestion of nutrients in the cell. In plants' cells, vacuoles are generally large. A Vacuole is a single layer of unit membrane enclosing fluid in a sack-like bubble. It produces turgor pressure (rigidity due to high pressure of water from the inside of a cell onto its membrane) to support the cell wall, Figure (17). It stores water and insoluble wastes of various chemicals.

Figure (17): vacule is pressurized with water to support the nucleus

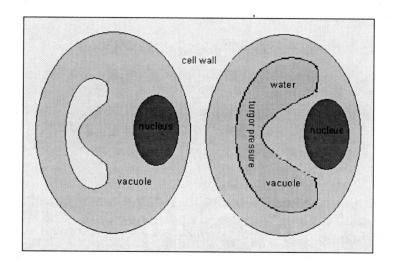

1-1-11 Centriotes within centrosomes

We said before that the cytoskeleton is used to differentiate cell polarity, the centrosome is a unique organelle that organizes microtubules to establish the cell polarity. Thus, the centrosome with the microtubules are the best organizing centers of the cell. The centrosome has three important activities, it::

- o nucleates the polymerization of tubulin subunits into longer polymers into microtubules
- o organizes the nucleated microtubules into useful arrays and matrices;
- o duplicates once every cell cycle

Each centrosome has two cylindrical centrioles that are perpendicular to each other. Each centriole has nine groups of microtubules, and each group has three microtubules as shown in Figure (18) below:

Figure (18): Centrosome and centrioles

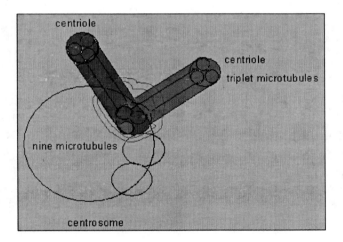

In the above figure, the two centrioles are linked together by fibers (red). Let's call the blue cenriole the mother and the green centriole the daughter. The centrosome is a complex organelle that contains about 100 proteins in addition to the γ-tubulin complex (protein needed to make up microtubules).

The Centrosome plays an important role in cell duplication in which the cell divides into two. The division of the cell happens in one cycle of the cell cycle. In normal division, both the centrosome and the chromosomal DNA must be duplicated only once per one cycle of the cell cycle. One cell cycle passes through four phases: G1, S G2, and M. In the G1 phase, there is one centrosome per cell that consists of two perpendicular centrioles. The duplication process of the centosome starts at the phase G1-S transition, at which the DNA also starts replication. In this phase (G1-S), the two perpendicular centrioles start to separate from each other. Once separated, centrioles start to grow orthogonally. At G2, each centriole becomes a pair of centrioles, Figure (19). This pair of centrioles has one new and one old centriole. In the duplication of somatic cells, there must be an old centriole to create a new centriole. This is not the case in all animals and plants where new centrioles are not of the same pattern.

Figure (19): Nuclear and centrosome cycle

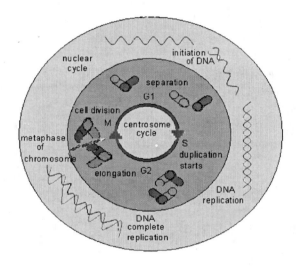

At the G2-M transition phase the duplicated centrosomes move to opposite direction of the nuclear envelope, Figure (19). At this phase, the new centrosomes meet the chromosomes and interact with them to create the bipolar spindle. This causes the chromosomes to separate. Recent research has shown that cyclin E and its associated E-Cdk2 are required for centrosome duplication, and this reaches its peak of activity at the G1-S transition. Also, the initiation of the DNA replication takes place at this stage.

1-2 Cellular respiration

Every time we take a breath, our blood transports oxygen to the mitochondria, where it is used to convert the nutrients in our food to a form of energy that the body can use. Problems with this process, which is called cellular respiration, have been linked to a number of morbid conditions, from unusual genetic diseases to diabetes, cancer and Parkinson's, as well as to the normal ageing process.

Cellular respiration is the process of oxidizing food molecules, like glucose, to carbon dioxide, water, and energy. The energy released is called adenosine triphosphate (ATP), and is used by all the energy-consuming activities of the cell.

The process occurs in two stages:

- o glycolysis: is the break down of glucose to pyruvic acid (one molecule of glucose breaks down into two molecules of pyruvate, which are then used to provide further energy,
- o pyruvate acid (CH3COCOH) is then oxidized to carbon dioxide and water The first stage occurs in the cytosol (cytoplasmic matrix), and the second stage occurs in the mitochondria.

One molecule of glucose breaks down into 2 molecules of pyruvate, or 6-carbon glucose converts to two of 3-carbon pyruvate. The pyruvic acid is oxidized by NAD^+ (Nicotinamide Adenine Dinucleotide) to produce $NADH + H^+$. The NAD consists of two nucleotides joined through their phosphate groups: with one nucleotide containing an adenosine ring, and the other containing nicotinamide. Adenosine is covalently bonded by two molecules of adenine and ribose molecules as shown in Figure (20) below.

Figure (20): Structure of adensine

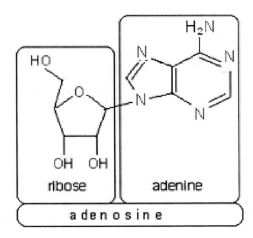

The Nicotinamide is known as niacinamide and is the amide of nicotinic acid (vitamin B_3), combined with ribose as shown in Figure (21) below.

Figure (21): Structure of nicotamide

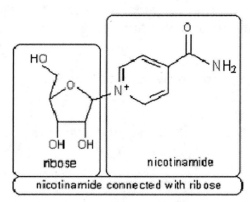

nicotinamide connected with ribose

Combining the above nicotinamide and adenosine, we get a molecule of NAD (Nicotinamide Adenine Dinucleotide), which is a coenzyme required by all living cells. The NAD has two nucleotides (dinucleotide) joined through two groups of phosphate, as shown in Figure (22) below.

Figure (22): Structure of NAD

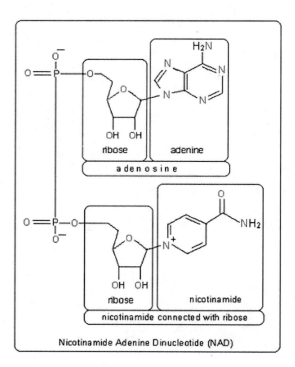

Nicotinamide Adenine Dinucleotide (NAD)

Now, we know the chemical formula (connection) of NAD, so we will move on to the abbreviation of ATP. ATP means adenosine triphosphate, and this is the energy released when the sugar is oxidized. The molecular formula of the ATP is as shown in Figure (23) below.

Figure (23): Structure of ATP

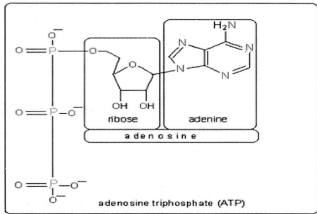

Similarly ADP (adenosine diphosphate) is shown in Figure (24) below.

Figure (24): Structure of ADP

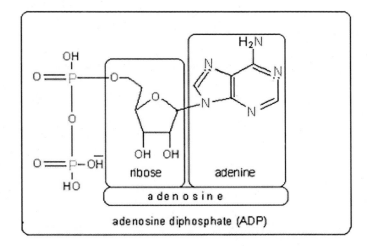

Adding water to the ATP will result in ADP, anions of phosphate, cation of hydrogen, and energy as in Figure (25).

Figure (25): ATP and ADP are interchangeable

The chemical structures show:

$$\text{Adenosine triphosphate (ATP)} + H_2O \rightleftharpoons \text{adenosine diphosphate (ADP)} + PO_4^{-3}\, 2H^+ + energy$$

Therefore,

$$ATP + H_2O \rightleftharpoons ADP + PO_4^{-3} + 2H^+ + energy$$

Lots of energy is released due to the hydrolisation of ATP. The produced ADP energy is used for any energy required in the cell for metabolic action. The Above equation can be put in a cycle form, Figure (25). This is called the ATP cycle, Figure (26).

Figure (26): ATP cycle, with $P = PO_4^{-3}$, see Figure (23), and Figure (24)

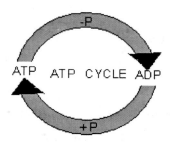

So, let's discuss the steps from the time our body takes the sugar to the time when we utilize the energy produced by the oxidation of the sugar:

1- Glucose of $C_6H_{12}O_6$ is oxidized and produces two molecules of pyruvate, water and energy. The pyruvate will be immediately combined with an acetyl enzyme called CoA, The acetyl is a radical molecule, and loves to combine with both the enzyme CoA and the pyruvate. The product of pyruvate and the acetyl CoA will produce citric acid.

2- Since there are two main carriers of energy in living cells, adenosine triphosphate (ATP) and nicotinamide adenine dinucleotide (NAD^+), pyruvate will be oxidized by NAD^+ to produce $NADH + H^+$.

3- Oxidation of the pyruvate makes it decarboxylated, producing CO_2.

4- An enzyme called CoA that assists in transferring fatty acids from the cytoplasm to mitochondria will meet some carbons produced due to the decarboxylation of pyruvates. Together with the double bond oxygen in the NAD, the CoA will be converted to acetyl-CoA, as the acetyl group has only one oxygen atom double bonded to a carbon atom.

5- The acetyl CoA will combine with the pyruvic acid that is composed through the glycolysis process of the glucose (sugar). The product is called oxaloacetic acid.

6- The oxaloacetic acid will combine with the citric acid that is involved in the physiological oxidation of fats, proteins, and carbohydrates to carbon dioxide and water (occurs in the metabolism of almost all living things). The result of the combination is the cis-aconitic acid; see flow chart below (or citric acid cycle, or kerb cycle). In effect, the cis-acid is the same as citric acid except that the latter is dehydrogenated, i.e., the oxaloacetic acid is a reducing agent to the citric acid. In other word, the citric acid is an acid, and the oxaloacetic acid is a base.

7- The cis-aconitic acid will regain the hydrogen again due to the introduction of NAD^+ (energy) that is still available from step 4 above. The cis-aconitic acid will become iso-cis aconitic acid and is called isocitric acid.

8- The isocitric acid will lose hydrogen molecules to become oxalsuccinic acid. The loss of hydrogen is because NAD^+ has been converted to NADH:

$$NAD^+ + H^+ 2e^- \longrightarrow NADH$$

9- In the glycolysis process of the glucose CO_2 is released. After the oxalsuccinic acid is formed, the CO_2 is released and another acid called the ketoglutaric acid is formed.

10- With the help of NAD^+, water and CO2 are released to produce scuccinyl CoA which will consequently release CO2 again to produce succinic acid.

11- Succunic acid will lose hydrogen molecule to become fumaric acid that will absorb water to convert to malic acid.

12- NADH will carry one hydrogen molecule to become oxaloacitic acid. The cycle will repeat itself in the process of glycolysis.

So, in the glycolysis process, the sugar of a molecular formula C6H12O6 will be converted to Carbon dioxide (CO2), water (H2O), and energy as the following equation:

$$C_6H_{12}O_6 + O_2 \longrightarrow CO_2 + H_2O + energy$$

The water in the above equation is not produced in one shot. The molecule of the citric acid in step 1 above has three double-bonded oxygens on the top of the molecule which makes it get rid of the water. The isocetric acid in step 8 has the opposite isocomponent of the citric acid and loves water. Therefore, the water lost with the citric acid is regained with the isocetric acid. The water in the citric acid cycle (Krebs Cycle) is released and regained and then released in the cycle. Carbon dioxide is released in two steps through the cycle, and the hydrogen atom is released four times. Thus, the glycolysis can be shaped with respect to water, carbon dioxide, and hydrogen as seen in the following diagram, Figure (27).

Figure (27): Glycolysis process

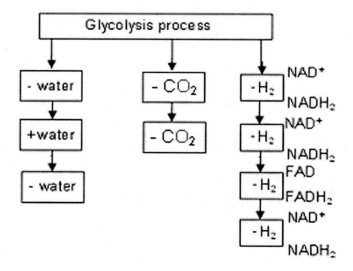

The complete citric acid cycle (Krebs cycle) is shown in Figure (28).

Figure (28): Complete citric acid cycle (Krebs cycle)

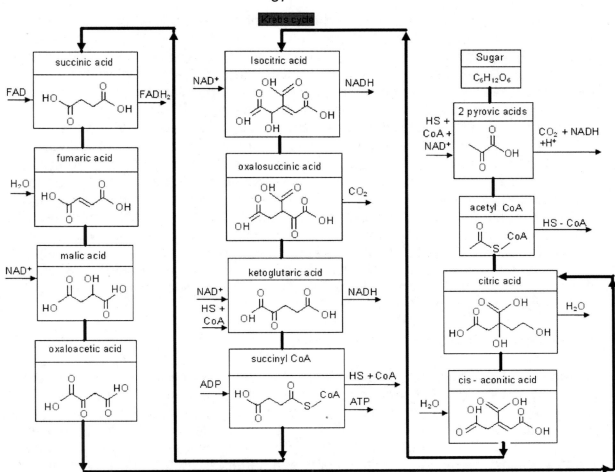

To summarize:

1- There are three interlinked energy production cycles:

      o the glycolitic (sugar burning)

      o the Krebs' citric acid cycle (protein and fat burnings)
      o electron transport

2- At the end of the glycolitic (pyruvic acid), electrons spark helping power the Krebs cycle to generate much of the ATP,

3- At the beginning of Krebs cycle, protein and fat need oxygen to burn them. This oxidation process is accomplished by the NAD which takes the hydrogen from foods to ease the oxidation. The NAD become oxidized, and the NADH is reduced (electron-energy rich).

4- The NADH is rich with electrons, and therefore generates much of the ATP energy that literally energizes our life. Each unit of NADH is capable of generating three units of ATP energy. When there is a lot of energy (ATP) in the body, it means that the food has lots of oxygen (or the food has been well oxidized). The ATP is generated due to the oxidative break down of pyruvate. The question is "can ATP be produced by the non-oxidative breakdown of the pyruvate?" This is the cause of CANCER as stipulated by Warburg's hypothesis which he won the noble prize for. Otto Warburg (1883) hypothesized that cancer cells generate energy (ATP) by non-oxidative breakdown of pyruvate(in the process that is called glycolisis). This is in contrast to "healthy" cells which mainly generate energy from oxidative breakdown of pyruvate. Pyruvate is an end-product of glycolysis, and is oxidized within the mitochondria. Therefore, according to Warburg, cancer should be interpreted as the dysfunction of the mitochondria. Warburg reported a fundamental difference between normal and cancerous cells to be the ratio of glycolysis to respiration; this observation is also known as the "Warburg effect". He quoted "the prime cause of cancer is the replacement of the respiration of oxygen in normal body cells by a fermentation of sugar". In fermentation the ATP produced is three units, whereas the ATP produced through the glycolisis process of normal cells is 36 units.

1-3 Molecular biology

Molecular biology covers a wide range of topics related to molecular and cell biology including functional and structural genomics, proteomics, molecular, enzemology, biomedicinal therapeutic, virology, immunology, biochemistry, nucleic acid, and other topics related to biology at a molecular level. Molecular biology concerns itself with understanding the interactions between the various organelles of a cell including the interactions between genetics, DNA, RNA, protein synthesis, and the controllability of these interactions. Molecular biology can be divided into four main topics:

1- Biochemistry is the study of actions and reactions of atoms and molecules occurring in living cells and organelles.
2- Genetics deals with cell division, alleles, genotypes, phenotypes and heredity.
3- Physiological metabolism is the transcription of RNA into proteins.
4- Microbiology is the science and study of microorganisms, including protozoan, algae, fungi, bacteria, and viruses.

Biochemistry deals with topics such as protein structure, enzyme mechanism, carbohydrate metabolism, lipids, and bioenergetics.

1-3-1 Protein structure

Proteins are responsible in part for maintaining functional stability and homeostasis of cells and tissues. During aging there are many chances for appropriately transcribed peptides and proteins from RNAs to become structurally altered.

Accumulation of altered proteins may be correlated with a loss of function or, in some cases, a gain of inappropriate or toxic functions. Therefore, it is critical to identify the specific alterations of proteins occurring during aging and in disease processes, and to define the role these alterations play in age-related and disease-related pathologies.

Proteins are chains (polymers) of amino acids. There are twenty amino acids that are commonly found in proteins. All amino acids (except proline) have similar structure. They have a carboxyl group, an amino group and a chiral (distinguishable) carbon (α carbon), Figure (29).

Figure (29): Molecular structure of amino acid

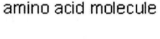

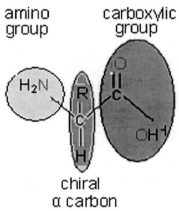

At physiological pH (pH = 7.3-7.4), the natural amino acids exist as zwitterions (two ions in German) with a positively charged amino group and negatively charged carboxyl group. Since the α carbon is chiral, the amino acid has an isomer configuration (stereoisomer) which differs in the space configuration as shown in Figure (30) below.

Figure (30): Two physical shapes (stereoisomers) of amino acid

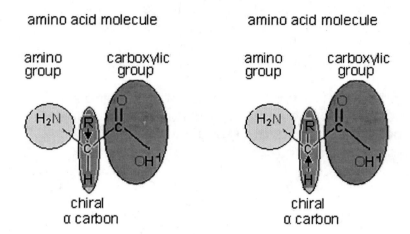

amino acid molecule   amino acid molecule

amino group   carboxylic group   amino group   carboxylic group

chiral α carbon   chiral α carbon

two different 3-dimention structures

The side chain or R group varies in all amino acids. The simplest chain is hydrogen in glycine amino acid. The tryptophan has cyclohexane of the aromatic smell. Amino acids have another important classification: whether they are polar (hydrophilic) or non-polar (hydrophobic), or whether they are an ionizable side chain (cations or anions). Table (1) shows the polar, non polar or ionized amino acids.

Table (1): Classification of amino acids

Amino acid	Polar	Non polar	Ionized
Table (1) Classification of amino acids			
Alanine		✓	
Cysteine		✓	✓
Glycine		✓	
Isoleucine		✓	
Leucine		✓	
Methionine		✓	
Phenylalanine		✓	
Proline		✓	
Tryptophan		✓	
Tyrosine		✓	✓
Valine		✓	
Arginine	✓		✓
Asparagine	✓		
Aspartate	✓		✓
glutamine	✓		
Glutamate	✓		✓
Hestidine	✓		✓
Lysine	✓		✓
Serine	✓		
threonine	✓		

Both the carboxyl group and the amino group of each amino acid are ionizable to become an acid or base, see Figure (31), depending on the pK of the environment the amino acid reacts in.

Figure (31): pK and combination of carboxyl group with amino group

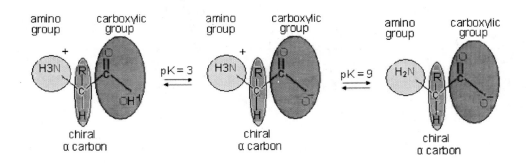

As an example, a tryptophan amino acid can go through three states of ionization; protonation, neutral, and deprotonation, based on the pH level of the media, Figure (32).

Figure (32): Three stages of tryptophan amino acid

Ionization of tryptophan amino acid

The configuration of the α amino acids is designated by the prefixed small capital letter D or L. For example D-glyceraldehydes and L-glyceraldehydes are the same except the OH and H are exchangeable as shown in Figure (33). The upper diagram is called Fischer model, and the lower is called the skeletal model.

Figure (33): L and D shapes and glycine

Proteins (peptides) are made of many amino acids joined together through the amide group and the carboxylic group. Any number of amino acids can be joined together to form protein (peptide) of any length. Polypeptide contain more than two dozens of amino acids, and when the polypeptide has less than a couple of dozen, the protein is called oligopeptide. Figure (34) shows a dipeptide and tripeptide.

Figure (34): Dipeptide and tripeptide

Notice that peptides have a polarity, a double bonded oxygen from the carboxyl group (hydrophilic), and a non-polar side (hydrophobic) on the amino group. The double polarity side (hydrophilic group) is called the backbone of the protein, and the other side is called the sidechain. All proteins have, as a minimum, two ionizable groups: the carboxyl group and the amino group. Some ions are also on the sidechain of the peptide (the peptide and the protein are the same if the number of amino acids are more than 24). It is important to be able to predict the charge and the type of the charge (positive or negative) carried by a particular protein for the purpose of electrophoresis (movement of particles under an electrical field), and chromatography (separation of proteins based on the colour). The charge of a protein depends on the concentration of the solution, or the pH and pK of the solution the protein resides in, i.e. the titrability of the solution.The charge of the protein is dependant on the pH of the solution. In other word, if the solution has a pH of 2, that means the solution is very acidic and concentrated with hydrogen, which makes the carboxylic side difficult to loose the hydrogen atom from the OH bond. Therefore, the protein is neutral or has some negative charge. If the solution is basic (say pH is 10), this means that the solution is hungry for the hydrogen, therefore, the protein tends to be negative, and so on. The total charge on a protein is simply the sum of charges of every ionized group in the protein. To find the net charge, identify the entire ionized group, determine the charge on each group at the specified pH, then add the

charges together, see Figure (35),
http://webhost.bridgew.edu/fgorga/proteins/default.htm

Figure (35): Net charge of four amino acids at pH 2, 7 and 11

net charge on this peptide at three different pH's (2, 7 and 11) is:

pH	charge on N-terminus	charge on C-terminus	net charge
2.0	1+	0	1+
7.0	1+	1-	0
11.0	0	1-	1-

1-3-2 Peptide and protein sequences

The primary structure (or sequence) of a peptide or protein is always written starting with the amino terminus on the left and progresses towards the carboxy terminus. If the entire sequence does not fit on one line it is simply continued on a second line, still following the left-to-right, amino-to-carboxy terminus convention. Amino acid sequences can be written using either the three letter code or a one letter code. The exact formatting of sequences varies with the application; by convention single letter codes are always capitalized.
Here are the orders of each amino acid.

Amino Acid	3 Letter Code	1 Letter Code	Amino Acid	3 Letter Code	1 Letter Code
Alanine	Ala	A	Leucine	Leu	L
Arginine	Arg	R	Lysine	Lys	K
Asparagine	Asn	N	Methionine	Met	M
Aspartate	Asp	D	Phenylalanine	Phe	F
Cysteine	Cys	C	Proline	Pro	P
Histidine	His	H	Serine	Ser	S
Isoleucine	Ile	I	Threonine	Thr	T
Glutamine	Gln	Q	Tryptophan	Trp	W
Glutamate	Glu	E	Tyrosine	Tyr	Y
Glycine	Gly	G	Valine	Val	V

Examples of primary structures:

Example	Three letter code	One Letter Code
A small peptide (8 residues)	AspIleGluPheArgValLeuHis	DIEFRVLH
Lysozyme* (129 residues)	LYS VAL PHE GLY ARG CYS GLU LEU ALA ALA ALA MET LYS ARG HIS GLY LEU ASP ASN TYR ARG GLY TYR SER LEU GLY ASN TRP VAL CYS ALA ALA LYS PHE GLU SER ASN PHE ASN THR GLN ALA THR ASN ARG ASN THR ASP GLY SER THR ASP TYR GLY ILE LEU GLN ILE ASN SER ARG TRP TRP CYS ASN ASP GLY ARG THR PRO GLY SER ARG ASN LEU CYS ASN ILE PRO CYS SER ALA LEU LEU SER SER ASP ILE THR ALA SER VAL ASN CYS ALA LYS LYS ILE VAL SER ASP GLY ASN GLY MET ASN ALA TRP VAL ALA TRP ARG ASN ARG CYS LYS GLY THR ASP	KVFGRCELAA AMKRHGLDNY RGYSLGNWVC AAKFESNFNT QATNRNTDGS TDYGILQINS RWWCNDGRTP GSRNLCNIPC SALLSSDITA SVNCAKKIVS DGNGMNAWVA WRNRCKGTDV QAWIRGCRL

*from chicken egg white

1-3-3 Lysozyme polypeptide

A Lysozyme is an enzyme, commonly referred to as the "body's own antibiotic" which can be obtained from the whites of chicken eggs. It plays a major role in the development of protein immunogenicity and antigenecity. It contains 129 amino acid residues in a single polypeptide chain, cross linked with four disulfide bridges. Lysozyme kills positive gram and negative gram bacteria (gram-positive cell walls consist of many layers. Peptidoglycan does not posses a lipid outer membrane. On the other hand, Gram-negative cell walls have only one or a few layers of peptidoglycan but posses an outer membrane consisting of various lipid complexes).

Figure (36): Computer image of lysozyme polypeptide

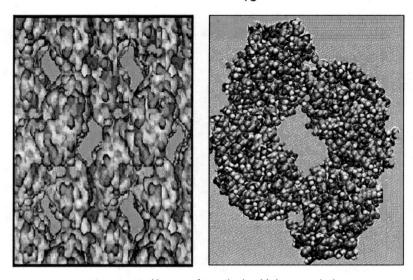

computer generated images of an orthorhombic lysozyme lattice

Lysozyme consists of 129 amino acids with 1001 non-hydrogen atoms

http://www.pubmedcentral.nih.gov/articlerender.fcgi?artid=2045120

The polypeptides (polymers) consist of a sequence of 20 different L-α-amino acids, also referred to as residues. To be able to perform their biological function, proteins fold into one or more specific spatial conformations, driven by a number of covalent and noncovalent interactions, such as hydrogen bonding, ionic interactions, hydrophilic grouping, and Van der Waals forces. A number of residues, not less than 40, are required to perform a functional protein. A more complete protein needs around 300 residues to perform multiple functions. Protein structure passes through four positions, Figure (37).

Figure (37): Four positions of protein structures

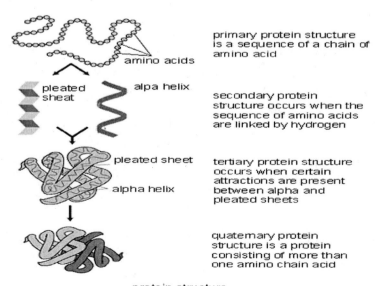

primary protein structure is a sequence of a chain of amino acid

secondary protein structure occurs when the sequence of amino acids are linked by hydrogen

tertiary protein structure occurs when certain attractions are present between alpha and pleated sheets

quaternary protein structure is a protein consisting of more than one amino chain acid

protein structure

http://commons.wikimedia.org/wiki/File:Protein-structure.png

These four positions are:

- o Primary structure; the sequence of amino acids arrangement.
- o Secondary structure; the diversion of primary to alpha helix and strands of pleated (beta) sheets.
- o Tertiary structure; three dimensional structure of a single protein molecule. It is the complete folded and compacted polypeptide chain.
- o Quaternary structure; it is the final position, which functions as a complicated protein complex.

A protein can go from one structure to another to perform the required biological function, and it can also go reversible.

1-4 Enzymes and enzyme mechanism

Proper enzyme intake improves digestion, elimination, and detoxification. With over 100 different enzymes in the body and all of them are important to an anti aging process.

Enzymes are proteins that act like catalysts by binding to an initial substance and converting it into something else. Because of this enzymes are needed for every chemical reaction that takes place in your body. No mineral, vitamin or hormone can do any significant work without enzymes. It is for this reason that enzymes are critical for your health and factor so importantly in an effective anti aging process.

Enzymes are biomolecules that increase (or decrease) the rate of chemical reactions. Almost all enzymes are proteins used to enhance reaction rates of substrates. For example, when adding the enzyme rennin in milk for making cheese, it causes the milk to coagulate. The milk is the substrate, and the rennin is the enzyme. The mechanism by which enzymes catalyze chemical reactions begins with the suitable match between the substrate and the enzyme (the active site). If the active site does not complement the substrate site, then the enzyme has no effect. This is called 'lock and key match", Figure (38).

Figure (38): Lock and Key match between enzymes and subtrates

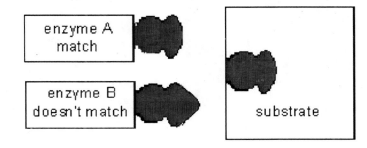

The enzyme can match the substrate when the following conditions are achieved:

- o Orientation and proximity
- o Electrostatic match
- o Preferential binding
- o Induced match
- o pH effect
- o nucleophilic bond
- o metal ionphilic (such as sulfur, phosphorous, calcium)

Proximity and orientation: proximity means the two molecules of enzyme and substrates are already in contact with each other to increase the local concentration of reactants. Orientation must be in optimal position to react. For example the carboxyl group of the enzyme would have a faster reaction if it is positioned towards the amino group of the substrate.

Electrostatic charges have effects on the polarity of the enzyme (Fajan's formulation). The reaction between enzymes and substrates is influenced by the exchange of charged atomic components. For example, a cation of an amino acid $(H_3N)^+$ would distance itself from an electrostatic charge of a positive magnitude, and therefore the reaction rate would be reduced.

Preferential binding is whether a specific enzyme reacts faster in a substrate containing K^+ (cation of potassium) or Na^+ (cation of sodium). The amino group of the enzyme may react faster with potassium or sodium, which both have one electron in the outermost orbit.

Induced dozes of enzymes could increase the reaction to a certain limit and in a defined pace of a reaction of a straight rate, exponential or steady rate. Figure (39) shows the rate of an enzyme induced into a substrate.

Figure (39): Effect of an induced enzyme onto reaction (some cases have different rate of reaction or a combination of an exponential and straight line).

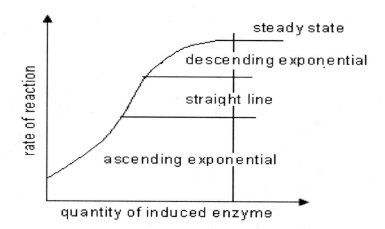

The pH effect of the reaction is different from one enzyme to another. The pH of the stomach is very low (pH=1), and its effect on the enzyme pepsin is optimum at a pH from 1 to 2. The pH of the blood in the cytoplasm (cytosol) is about 7, and has an optimum effect on the enzyme carbonic anhydrase which is required to convert carbon dioxide into bicarbonate and hydrogen, Figure (40). See Chapter 3 for the effect of pH on aging process.

Figure (40): Different optimum reaction for different pH levels
http://academic.brooklyn.cuny.edu/biology/bio4fv/page/ph_and_.htm

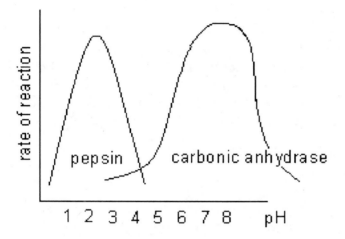

The pH of a substrate can have several effects on the shape and activity of the enzyme. If the ionization state (acidic in the carboxyl group and basic in the amino group) of the enzyme is altered due to the level of the pH in the substrate, the shape of the enzyme can be altered and consequently the orientation and proximity will be changed. This can change the enzyme recognition and its activity of reaction.

1-5 Nucleophilic bond

A nucleophile is a nucleus lover and is opposite to an electrophile, which is an electron lover. A nucleophile and an elecrophile form a chemical bond by donating and accepting electrons respectively. By the definition of Lewis dot, nucleophiles donate electrons, therefore, all enzymes with free electrons (anions) can act as nucleophiles, and those enzymes with free protons (cations) can act as electrophiles. In other words, a nucleophilic match can happen when the enzyme bonds with a hydrogen such as NAD^+, which bonds with hydrogen (NADH).The result is another product, Figure (41).

Figure (41) Nucleophic NAD+ bonds with hydrogen (proton)

$$\underset{\text{L- lactate}}{\overset{\displaystyle COO^-}{HO-\overset{\displaystyle |}{\underset{\displaystyle |}{C}}-H}} + NAD^+ \underset{\xleftarrow{\hspace{1cm}}}{\overset{\text{lactate dehydrogenase}}{\xrightarrow{\hspace{1cm}}}} \underset{\text{pyruvate}}{\overset{\displaystyle COO^-}{HO-\overset{\displaystyle |}{\underset{\displaystyle |}{C}}=O}} + NADH + H^+$$

1-5-1 Metal Ionphilic

Metal binding locations in enzymes and proteins are varied in their geometries, metal preference, and ligands which include backbones and sidechains. For example, iron (Fe) bonds to the sulfur (S) of enzymes or protein from the sidechains of cysteine and methionine residues, whereas some metals bond from the backbones, and some bond at the centre of the enzyme (at the hydrophilic ligands).

1-6 Bioenergetics

Bioenergetics concern energy flow through living systems during respiration and metabolic processes. An important group of enzymes including myokinase, creatine kinase, and the nucleoside diphosphate kinase exist between the inner and the outer membrane of the mitochondrion. Myokinase is an acid-stable protein occurring in the skeletal muscle; when acting with enzyme hexokinase:

Adenosine diphospahte + hexose $\longrightarrow$ adenylic acid + hexose monophosphate.

Chemical formula of the equation is shown in Figure (42).

Figure (42): Product of ADP and hexonase

adenosine diphosphate (ADP)   hexose (monosaccharide)

adenylic acid (adenosine monophosphate, AMP)   hexose monophosphate

The myokinase can have the same job of transphosphorylation as the hexose but it works with the ATP (adenosine triphosphate). So,

ATP + myokinase ⟶ ADP + hexose monophosphate

The mechanism of the transphosphosphrylation by the myokinase has so far not been understood. However, Dr. M. Johnson of the university of Wisconsin, suggested thatmyokinase might catalyze a reversible transfer of phosphate from adenosine diphosphate to another, yielding adenosine triphosphate and adenylic acid as follows:

2 ADP + myokinase ⟶ ATP + adenylic acid (AMP)

Another enzyme which is called creatinekinase (CK), also known creatine phosphokinase (CPK) has an important effect on reducing energy off muscles, and released by various tissues of the body. It consumes the ATP and the creatine to create ADP and phosphocretine (PCr):

CPK + ATP ⟶ ADP + phosphocreatine

The energy flow in the body is complicated to visualize, but we shall try to simplify it in a better way. Skeletal muscles need the ATP to synthesize protein and help with substitution of the energy lost needed for sustaining the metabolism. The ATP is not available in sufficient quantity during intense periods. The ATP is produced primarily from the burning of sugar, glycogen stored in the liver, and fatty acids. If this ATP is not sufficient, then the phosphate group of phosphate creatine (PCr) liberated by the effect of the enzyme CPK, is then bound to the ADP, resulting in the production of ATP and free creatine (Cr) (Hochachka, 1994), as per the flow chart of Figure (43).

Figure (43): Generation of ATP (1) from sugar and lipid, (2) from glycogen, and (3) form creatine phosphokinase (CPK)

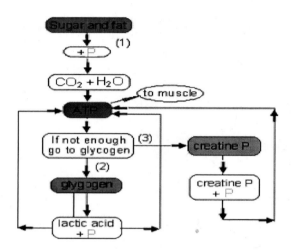

The question one can ask is, why is phosphorous, and not other atoms (such as sodium, magnesium, sulfur, or others) involved in the energy? The answer is simple: a- phosphorous is available in the body in large quantities, and b- the first ionization energy (kJ/mol), is the largest energy. For example:

Sodium has 496 kJ/mol, magnesium 736, potassium 419, sulfur 1000, phosphorus 1012.

1-7 Cell divisions

Cells from older individuals generally do proliferate at a slower rate than those from younger individuals when grown in culture. The difference is not usually very dramatic, however, perhaps half the rate of cells from a newborn infant at most. This evidence would indicate that the phenomenon is not dependent upon nutrition, since these cells would be grown under identical conditions. Do cells devide at a faster rate in meiosis than in mitosis? Our current knowledge from cells in culture indicates that unless they are "transformed" or "immortalized", as occurs in cancer cells, they all have a finite life span and have an effect on aging.

1-7-1 Mitosis

Eukaryotic cells reproduce in two different ways: mitosis and meiosis. In mitosis, the nucleus divides into two nuclei (referred to as karyokinesis), and the cytoplasm divides into two portions (referred to as Cytokinesis), Figure (44).

Figure (44): Reproduction of two cells from one cell (mitosis)

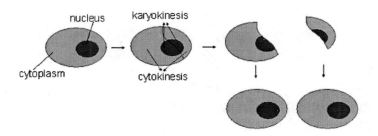

In the karyokinesis division, the chromosomes separate first and then are followed immediately by cytokinosis, in which the cell divides in to two identical daughters, Figure (45).

Figure (45): Division of nucleus

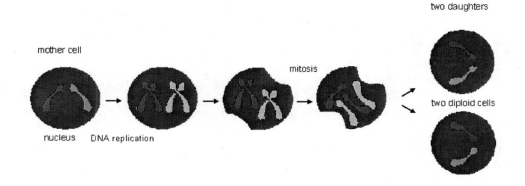

The period between successive divisions is known as interphase. The interphase period is divided into a number of stages, Figure (45a).

Figure (45a): stages of cell division (mitosis)

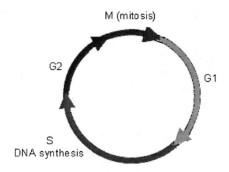

Some cells divide rapidly. For example, beans take about 19 hrs to complete the cycle division, whereas red blood cells divide at a rate of 2.5 million per second. Cancer cells divide rapidly, the daughter cells divide before they have reached maturity. Electrocharge, pH, temperature, and some drugs (enzymes) may affect the rate of division. When cells stop dividing, they stop at a point late in the stage G1. The stage S is the stage when the DNA is replicated for the next division, and the chromosomes become double stranded. The cell has then entered the G2 stage and proceeds in to cell division. Cells will not divide again and stop in the G1. The process of division passes into the following steps:

1-7-1-1 Prophase

Prophase is the first step of mitosis preparation where the cell is about to divide, and the chromosomes become visible and start to condense into double stranded chromosomes, Figure (46). Chromatin/DNA do not replicate in this step. Gradually, a spindle composed of protein fibers together with kinetochores forms and extends nearly the length of the cell, expanded in its centre (or the equator of the cell) like a base ball. The chromosomes each consist of two chromotids attached to each other by a spindle fiber at the centromere.

Figure (46): Double stranded chromosome divides into two identical sisters.

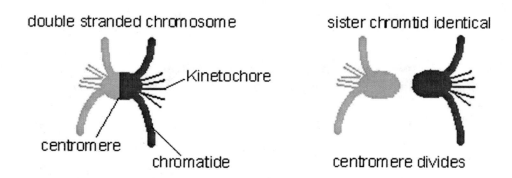

1-7-1-2 Metaphase

When the centomere and kinetochore arrive at the centre of the cell (equator), the metaphase begins. The mitosis process ends when the centromere divides so that each of the chromatids becomes a single stranded chromosome, Figure (46).

1-7-1-3 Anaphase

During anaphase, each sister of a single-stranded chromosome moves towards one pole of the cell (one sister to one pole, and the other to the opposite pole). Thus, anaphase is the stage when sisters of chromosome migrate to the two poles of the cell, Figure (47).

Figure (47): Movement of daughter chromosomes from equator to poles during anaphase

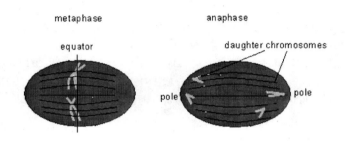

1-7-1-4 Telophase

In telophase, the nuclear envelope (which had disappeared during prophase) reforms, and chromosomes uncoil into the chromatin form again. There are now two cells instead of one cell, but they are of smaller sizes of the same genetic heredity. The small cells will develop into mature ones.

1-7-1-5 Cytokinesis

Cytokinesis is the last stage of cell division, where the daughter cells split apart. The mitosis is the division of the nucleus, and the cytokinesis is the division of the cytoplasm.

1-7-2 Meiosis

During meiosis chromosomes are duplicated once in S stage. Meiosis has two stages; the meiosis 1 is just like mitosis where the cell divides once. In meiosis 2 the cell divides again and the final product is four daughter cells. The process of meiosis is as follows:

- The first step in meiosis is the diploids of each of the 46 chromosomes duplicated, Figure (48). This is called DNA replication.
- The homologues of chromosomes (23 homologues in 46 chromosomes; i.e. one chromosome has two homologues) are segregated in two cells.
- Homologues pair up alongside, lengthwise, no.4 of Figure (48).
- The pairs of homologues are shuffling and cross over in metaphase 1, figure (48).
- At the end of meiosis, new chromosomes will be created to become part of eggs and sperms, which will be unique to their parent.
- During meiosis 2, the sister chromatids of each of the 23 chromosomes are pulled apart resulting in 4 cells. Each of the 23 chromosomes are now haploid.
- The Meiosis process is the way to create the diversity of all sexually reproducing organisms.

The invention of electronic microscopes allowed biologists to discover the basic facts of cell division and sexual reproduction. The center of genetics and heredity research then shifted to understanding what really happens in the transmission of hereditary traits from parents to children. A number of hypotheses were suggested by researchers but Gregor Mendel, a monk from Austria was the father of heredity.

1-8 Heredity

Genetics is the branch of biology that deals with heredity (family likeness). Heredity is the passing of characteristics (traits) from parents to offspring. The genetics of aging is the science of heredity for traits related to aging such as: lifespan, age at menopause, age at onset of specific diseases in late life (Alzheimer's disease, prostate cancer, etc.), rate of aging (estimated through tests for biological age), rate-of-change traits, and biomarkers of aging (Arking, 1998). In practice, most studies are focused on lifespan, because other reliable markers of aging are lacking or less convenient to use. Therefore, the genetics of aging is closely related to the biology of lifespan (Gavrilov & Gavrilova), see Chapter 3.

Cells can be different from each other because each cell or group of cells are specialized in particular functions such as sensing light (eyes' cells), fighting disease (immune cells), carrying oxygen (red blood cells), hair colour, etc. One cell has a nucleus in the centre that has 46 chromosomes. The 46 chromosomes have more than 35,000 genes. Each chromosome is packed with DNA, genes, proteins, and other kinds of molecules, Figure (49).

Figure (49): Chromosome, DNA, and histone

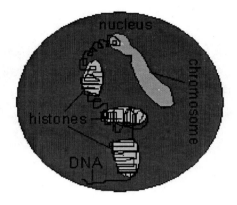

The name chromosome comes from the Greek word (chromo) means colour, and (soma) which means body. Chromosome means uniqueness of a human. Chromosomes are thread-like structures located inside the nucleus of animal and plant cells. Each chromosome is made of DNA (deoxyribonucleic acid) that is passed from parents to their offspring. Chromosomes vary in number and shape among living things. Humans have 46 chromosomes (23 pairs), where most bacteria have one or two circular chromosomes. For example, a fruit fly has four pairs, and a dog has 39. Some offsprings inherit their traits from their mother and others from their father. Only mothers keep chromosomes of their mitochondria during the fertilization of egg cells. Females have two X chromosomes in their cells, and males have one X and one Y chromosome. Inheriting imbalance of chromosomes may lead to health problems. For example, women with only one X chromosome are short and have some heart and kidney problems. Men who have more X chromosomes than Y have tall stature.

One cell has twenty two autosomes (chromosomes divide through the mitosis process and are the same in both sexes in humans. They have been used in genetic testing for ancestry, ethnicity, race, and genealogical purposes), and one sex chromosome. The longest chromosome is referred to as chromosome 1, the next longest as chromosome 2, and so on to the smallest autosome, chromosome 22.

In his experiments on peas Mendel considered plants traits such as colour, tall, shape and other visible traits. He did not account for hidden characteristics and traits such as taste, emotion, intelligence, age, and other factors dominating animal behaviour. Offspring may or may not be identical physical copies of parents; rather they are the combination of all characteristics and traits whether visible or invisible, sensible or insensible. Let's take, for example the first law of Mendel, Table (2).

Table (2): Genotypes crossing

	Y	y
Y	YY	Yy

y	Yy	**yy**

Three genotypes have traits influenced by the dominant phenotype in parents. If Y represents the tall, and y represent the short, then three genotypes are tall, and one is short. Expand the above behaviours to include two traits in each parent:

Consider:

Y: dominant allele for tall
y: recessive allele for short
W: Dominant allele for white
w: recessive allele for black

The output, as per Mendel would be combination of YyWw x YyWw, Table(3).

Table (3): Crossing between dominant and recessive alleles

	YW	Yw	yW	yw
YW	YWYW (tall)	YWYw (tall)	YWyW (tall)	YWyw (tall)
Yw	YwYW (tall)	YwYw (tall)	YwyW (tall)	Ywyw (tall)
yW	yWYW (tall)	yWYw (tall)	**yWyW (short)**	**yWyw (short)**
yw	ywYW (tall)	ywYw (tall)	**ywyW (short)**	**Ywyw (short)**

Note that there are twelve tall alleles, and only four shorts alleles.
For simplicity, alphabetical and numerical figures will be used.
Let us expand the alleles to include a third character dwarf and brown. Table (4) would be a combination of ABC123, and still the dominant is A in the father and 1 in the mother, Table (4).

Table (4): Expanding traits

	A1	A2	A3	B1	B2	B3	C1	C2	C3
A1	A1A1	A1A2	A1A3	A1B1	A1B2	A1B3	A1C1	A1C2	A1C3
A2	A2A1	A2A2	A2A3	A2B1	A2B2	A2B3	A2C1	A2C2	A2C3
A3	A3A1	A3A2	A3A3	A3B1	A3B2	A3B3	A3C1	A3C2	A3C3
B1	B1A1	B1A2	B1A3	B1B1	B1B2	B1B3	B1C1	B1C2	B1C3
B2	B2A1	B2A2	B2A3	B2B1	**B2B2**	**B2B3**	B2C1	**B2C2**	**B2C3**
B3	B3A1	B3A2	B3A3	B3B1	**B3B2**	**B3B3**	B3C1	**B3C2**	**B3C3**
C1	C1A1	C1A2	C1A3	C1B1	C1B2	C1B3	C1C1	C1C2	C1C3
C2	C2A1	C2A2	C2A3	C2B1	**C2B2**	**C2B3**	C2C1	**C2C2**	**C2C3**
C3	C3A1	C3A2	C3A3	C3B1	**C3B2**	**C3B3**	C3C1	**C3C2**	**C3C3**

The recessive of the genotype is 16 out of a population of 81. If there are two dominants in the father, and two dominant in the mother, the result would be one recessive trait and 80 dominants traits, Table (5).

Table (5): Two dominants in each of the parent.

	A1	A2	A3	B1	B2	B3	C1	C2	C3
A1	A1A1	A1A2	A1A3	A1B1	A1B2	A1B3	A1C1	A1C2	A1C3
A2	A2A1	A2A2	A2A3	A2B1	A2B2	A2B3	A2C1	A2C2	A2C3
A3	A3A1	A3A2	A3A3	A3B1	A3B2	A3B3	A3C1	A3C2	A3C3
B1	B1A1	B1A2	B1A3	B1B1	B1B2	B1B3	B1C1	B1C2	B1C3
B2	B2A1	B2A2	B2A3	B2B1	B2B2	B2B3	B2C1	B2C2	B2C3
B3	B3A1	B3A2	B3A3	B3B1	B3B2	B3B3	B3C1	B3C2	B3C3
C1	C1A1	C1A2	C1A3	C1B1	C1B2	C1B3	C1C1	C1C2	C1C3
C2	C2A1	C2A2	C2A3	C2B1	C2B2	C2B3	C2C1	C2C2	C2C3
C3	C3A1	C3A2	C3A3	C3B1	C3B2	C3B3	C3C1	C3C2	**C3C3**

Only one recessive is in 81 populations.

The number of dominants follows the equation:

$$D = (n_f \cdot n_m)^2 - ((n_f - n_{fd}) \cdot (n_m - n_{md}))^2 \qquad (1)$$

Where D is the number of dominants.
     n_f is the number of alleles in father,
     n_m is the number of alleles in mother,
     n_{fd} is the number of dominants in father,
     n_{md} is the number of dominants in mother.

Maximizing equation (1) by differentiating and making it equal to zero, it follows that the larger the number of dominant genotypes is the smaller number of recessive ones.

If the mother has no dominant, then equation (1) will depend only on the crossing between the number of alleles in both parents and the dominant in the father.

Equation (1) is equivalent to the Hardy-Weinberg equilibrium which is $p^2 + 2pq + q^2 = 100\%$. Equating both equations, one can get:

$$(n_f \cdot n_m)^2 = 100\% \qquad (2)$$
$$D = p^2 + 2pq \qquad (3)$$
$$((n_f - n_{fd}) \cdot (n_m - n_{md}))^2 = q^2 \qquad (4)$$

And therefore:

$$(p^2 + 2pq)/ q^2 = (n_f \cdot n_m)^2 - ((n_f - n_{fd}) \cdot (n_m - n_{md}))^2 / ((n_f - n_{fd}) \cdot (n_m - n_{md}))^2 \qquad (5)$$

If the number of dominant alleles in the father or mother equals the number of traits, the denominator will be equal to zero, and therefore the whole population will be dominants.

Now consider a man and a woman from the Amazon forest married to each other; their dominant phenotypes are intact and undamaged, due to the environment and the natural food, which does not contain chemicals or pollutants. The product of the genotype population will all be dominant. However, if their first offspring is married to a tall, white man from Greenland, their offspring would probably be a mixture between dominant and recessive traits. But how can this offspring from the Amazon get married to that offspring from Greenland, without overcoming transportations, obstacles and other related barriers? The question aroused is that 'if no change' in the environment takes place and there is no change in social behaviour, would genotypes have recessive traits. In other words, is there any evolution as per Darwinian theory, if there is no change in the natural world? This is a simple way to prove that natural selection is mathematically impossible for producing evolutionary change, because there is no gene drift when Amazonians marry each other.

To prove that Darwin theory can not be accepted, let us assume that the whole population is married to a father of pure recessive alleles, for the worst case scenario, as shown in Table (6).

Table (6): Equal recessive genotypes of a father of pure recessive alleles

		A1	A2	A3	B1	B2	B3	C1	C2	C3
B2		A1	A1A1	A1A2	A1A3	A1B1	A1B2	A1B3	A1C1	A1C2	A1C3
B3		A2	A2A1	A2A2	A2A3	A2B1	A2B2	A2B3	A2C1	A2C2	A2C3
B4		A3	A3A1	A3A2	A3A3	A3B1	A3B2	A3B3	A3C1	A3C2	A3C3
C2		B1	B1A1	B1A2	B1A3	B1B1	B1B2	B1B3	B1C1	B1C2	B1C3
C3	x	B2	B2A1	B2A2	B2A3	B2B1	B2B2	B2B3	B2C1	B2C2	B2C3
C4		B3	B3A1	B3A2	B3A3	B3B1	B3B2	B3B3	B3C1	B3C2	B3C3
D2		C1	C1A1	C1A2	C1A3	C1B1	C1B2	C1B3	C1C1	C1C2	C1C3
D3		C2	C2A1	C2A2	C2A3	C2B1	C2B2	C2B3	C2C1	C2C2	C2C3
D4		C3	C3A1	C3A2	C3A3	C3B1	C3B2	C3B3	C3C1	C3C2	C3C3

equals
to

A1A1R	A1A2R	A1A3R	A1B1R	A1B2R	A1B3R	A1C1R	A1C2R	A1C3R
A2A1R	A2A2R	A2A3R	A2B1R	A2B2R	A2B3R	A2C1R	A2C2R	A2C3R
A3A1R	A3A2R	A3A3R	A3B1R	A3B2R	A3B3R	A3C1R	A3C2R	A3C3R
B1A1R	B1A2R	B1A3R	B1B1R	B1B2R	B1B3R	B1C1R	B1C2R	B1C3R
B2A1R	B2A2R	B2A3R	B2B1R	B2B2R	B2B3R	B2C1R	B2C2R	B2C3R
B3A1R	B3A2R	B3A3R	B3B1R	B3B2R	B3B3R	B3C1R	B3C2R	B3C3R
C1A1R	C1A2R	C1A3R	C1B1R	C1B2R	C1B3R	C1C1R	C1C2R	C1C3R
C2A1R	C2A2R	C2A3R	C2B1R	C2B2R	C2B3R	C2C1R	C2C2R	C2C3R
C3A1R	C3A2R	C3A3R	C3B1R	C3B2R	C3B3R	C3C1R	C3C2R	C3C3R

One can see that, as per Mendel's laws, the number of dominant genotypes is still the same. In the table above, there are even more recessive components that have been added in the favour of the Darwin theory. Also, if same the recessive components are added, the number of dominants will be increased as shown in equation (5) if it is multiplied by q:

$$q \cdot (p^2 + 2pq)/ q^2 = q \cdot ((n_f \cdot n_m)^2 - ((n_f - n_{fd}) \cdot (n_m - n_{md})))^2 / ((n_f - n_{fd}) \cdot (n_m - n_{md}))^2$$

$$= (p^2 + 2pq)/q$$

The denominator will be q instead of q^2, and therefore the percentage of dominant to recessive will be larger than before.

Inevitably, the calculations are suggestive of the genetic consistency within a given population if there are no intercultural marriages. Furthermore, environmental changes and cross-cultural engagements have only recently been compared to the length of time: 3 - 7 million years it takes for evolution to take place according to evolutionists, http://www.drelser.com/shop/article_2/Heredity%2C-Darwin-and-misleading-theory.html?shop_param=cid%3D1%26aid%3D2%26

Nucleic acid and genes

Nucleic acids (DNA and RNA) specify and aid the construction of proteins which provide the functional elements of biological systems. DNA stands for deoxyribonucleic acid and RNA stands for ribonucleic acid, both of which are composed of nucleotides, Figure (50).

Figure (50): DNA and RNA are composed of sugar, phosphate and nucleotide.

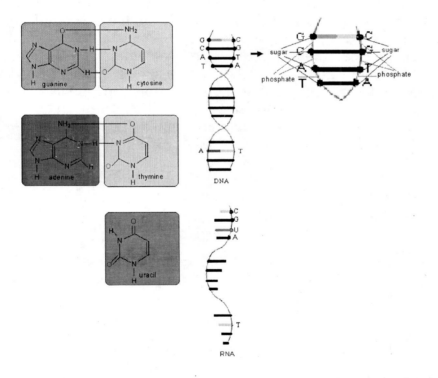

In our body, there are hundreds of thousands of proteins. These proteins are made locally. An enormous amount of information is required to manufacture them according to specific codes transmitted from the DNA to the exact location by the RNA.

The information required for designing each protein is stored in a set of molecules called nucleic acids. There are two types of nucleic acids: DNA and RNA. Both of them are made up of very large molecules that have two main parts. The nucleic acid has the backbone, which is made of alternating sugar and phosphate molecules bonded together in a long chain, as seen in Figure (50). Sugars in the backbones are connected to nucleotide bases called adenine (A), cytosine (C), guanine (G), and thymine (T). The DNA has two backbones, connected together by two nucleotide bases such that C and G is one bond, and A and T is another. Bonds are connected in alternate ways. Each nucleic acid contains millions of nucleotide bases. They are arranged in different groups of bonds so that each group represents a certain code of information carried to each specific protein with the same information as the genetic traits of parents. Each group is an exact genetic alphabet on which each of our proteins is coded. The two strands of the DNA wrap around each other forming a coil, or helix. The DNA has the ability to copy itself and self-replicate when needed during cell division, cell growth, and DNA repair. Copying or replication can be achieved in two steps. In the first step the hydrogen bonds between the nucleotide bases break and the two strands separate. In the second new complementary bases are brought and paired up with the same bases of the DNA, thus forming identical DNA.

Ribonucleic acids (RNA) is similar to DNA in construction, except that RNA does not have thymine. Instead it has uracil (U), and has only one single strand of backbone with similar construction to DNA (sugar and phosphate). RNA is used as transporter and messenger of information from the DNA out of the cell to help manufacture protein.

There are mainly three types of RNA: messenger RNA (mRNA), transfer RNA (tRNA), and ribosomal RNA (rRNA). Each one of them has a special function. mRNA contains information on the sequence of amino acids stored in the DNA, and carries the genetic code from the DNA to the ribosomes. The rRNA is in the cytoplasm combines with protein to make the ribisomes. The tRNA bonds to amino acids and then synthesizes the protein, Figure (51).

Figure (51): mRNA takes information on amino acids and equips the ribosome to manufacture the specific protein when it receives the amino acids that are delivered by tRNA.

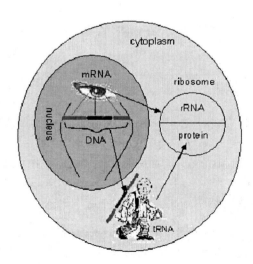

Transfer tRNA is the information adapter (like electrical connector) molecule. It is the direct interface between amino-acid sequence of a protein and the information in DNA.

Messenger or mRNA is a copy of the information carried by a gene on the DNA. The role of mRNA is to move the information contained in DNA to the translation machinery (tRNA).

Ribosomal RNA (rRNA) is a component of the ribosomes (sugar) and the protein synthetic factories in the cell

The double helix of the DNA in detail is shown in Figure (52). The figure shows that the backbones are composed of sugar and phosphate. The two backbones look like a spiral stair case, and connect to bases of adenine (A), thymine (T), cytosine (C) and guanine (G). A and T are connected by two hydrogen atoms, and G and C by three hydrogen atoms.

Figure (52): DNA molecule with two views

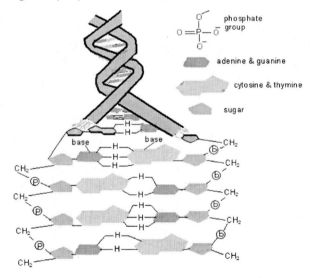

1-9 Immunology

Aging is not synonymous with illness. However, getting older does increase the risk for many diseases and disorders. Overall, elderly people have an increased rate of chronic disorders, arteriosclerosis ("hardening of the arteries"), infections, autoimmune disorders, and cancer.
The thymus, one of the organs of the immune system, is the site where certain immune cells called T lymphocytes ("T cells") mature. The thymus begins to shrink (atrophy) after adolescence. By middle age it is only about 15% of its maximum

size. Although the number of T cells does not decrease with aging, T cell function decreases. This causes a weakening of the parts of the immune system controlled by these T cells.

Immunology is a complex group of cells and organs that defends the body against disease and infection. It is the branch of biology concerned with the structure and the function of the immune system. Immunity is the distinction between self and non-self. We shall discuss the following subjects under immunology:

- o   Concepts of immunology
- o   Immune response
- o   Immune effector mechanism
- o   Immunity and health
- o   Tools and techniques in immunology

1-9-1 Concept of immunology

When a foreign pathogen or tumor cell invades our body, the immune system will respond in two ways; immediately and gradually. Immune cells capture and treat foreign cells in a hostile manner. Cells called the natural killers (NK) are among the first cells of the immune system that arrive at the site of infection or inflammation and kill the invaders. The immune system is a very complex subject in its function and structure. The interaction between innate immunity and adaptive immunity needs to be covered.

1-9-2 Innate immunity

This is the immunity mechanism that one is born with it is the initial response by the body to eliminate microbes and prevent infection. Innate immunity is the basic resistance and the first line of defense against infection and inflammation. The major function of the innate immunity response includes:

- o   Recruiting immune cells to the site of invasion, through the production of cytokins (cells with transmitters to call other immune cells for fighting).
- o   Activation of white blood cells.
- o   Activation of the adaptive immune system.
- o   Activation of the complement cascade to help clear pathogens and bacteria.

The innate immune response has some distinct characteristics such as:

- o   No memory to the invaders
- o   Responses are broad-spectrum, i.e. non-specific
- o   Limited repertoire of recognition molecules
- o   Responses are phylogenetically ancient

The innate immune system is responsible for destroying the majority of microorganism within minutes or hours.

1-9-3 Cells of the innate immune system

1-9-3-1 White Blood Cells (WBC)

White blood cells, known as leukocytes, are able to move freely and interact with and capture foreign objects and invading microorganisms. Leukocytes, unlike many other cells, are not tightly associated with organs or tissues, thus they move freely. WBCs cannot divide or reproduce on their own; They are produced in the bone marrow. Leukocytes include natural killer cells, mast cells, eosinophils, basophiles, and phagocytic cells, which include macrophage, neurophils, and dendrite cells.

1-9-3-2 Mast cells

Mast cells are found in tissues throughout the body on blood vessels, nerves, and in proximity to surfaces that interface with the external environment. They are produced in the bone marrow and depend on stem cells for their survival. Mast cells may be activated by the synthesis of cytokines and chemokines. They rapidly release characteristic granules rich in histamine and heparin along with other hormonal mediators. Histamine dilates blood vessels, causing signs of inflammation, and recruits neutrophils and macrophages. Mast cells appear to have a vital defense mechanism against parasitic infestations, immunomodulation of the immune system, tissue repair and angiogenesis.

1-9-3-3 Phagocytes

Phagocytes involve in the ingestion of pathogenic microorganisms. The plasma membrane expands around the particulate objects and form large vesicles called phagosomes, that will fuse with lysosomes. Both ingest the pathogenic microorganism. As mentioned before, phagocytes include monocytes, macrophages, neutrophils, and denritic cells.

1-9-3-4 Macrophages

A macrophage (macro means large, and phage means cell eating) is a type of white blood cell that ingests foreign microorganism; Macrophages are key player in the immune response to foreign invaders, such as infectious microorganisms. Blood monocytes migrate into the tissues of the body and there they differentiate, evolve into macrophages. Macrophages help ingest bacteria, protoza, and tumor cells. The binding of the microorganism cell to the receptors on the surface of a microphage triggers it to engulf and destroy the microorganism through the generation of a respiratory burst. This causes the release of reactive oxygen species that kill the microorganism.

1-9-3-5 Neutrophils

Neutrophils are the most common type of white blood cell. They comprise about 50 to 70 Percent of all blood cells. The bone marrow of a normal healthy adult produces more than 100 billion neutrophils per day, and more than 10 times that many per day during severe inflammation.They are the first immune cells to arrive at the site of invasion through a process known as chemotaxis (direction of microorganism movement towards food, for example sugar). Neutrophils die immediately after they ingest pathogens, and are short lived. They are fast acting, arriving at the site of invasion within an hour. When they attack the microorganism, they build a net of fibres called a neutrophil extracellular trap, which traps and kills invaders. They release complex proteins that help kill the invaders. They release chlorine bleach, which plays a part in killing microbes. A lowered neutrophil count results in a poor and weak immune system, and could cause anemia and leukemia. Patients with anemia have a count less than 1. The normal is between 2 and 7.

1-9-3-6 Dendritic cells

Dendritic cells (D.Cs) are present in skin, mucosal and respiratory membrane tissues, which are that is the first line of defense. They detect bacteria and viruses. Dendritic cells receive pathogens at their receptors of a toll-like shape, and phagocytose them through changing and processing the protein of such pathogens. Then the processed protein is carried to the T cells (we shall discuss T cells in the adaptive immunity section). Thus dendritic cells activate helper T cells, cytotoxic T cells and also B cells (adaptive immunity).

1-9-3-7 Basophils and Eosinophils

Basophils, eosinophils, mast cells, and immunoglubine E (IgE) constitute important elements in allergic inflammation. IgE is a class of antibody that has only been found in mammals. It plays an important role in allergies, and is especially associated with type 1 hypersensitivity. It may be important during immune defense against certain protozoan parasites. IgE is synthesized in response to allergens in the environment and become fixed to high-affinity receptors of basophiles, eosinophils and mast cells. These cells with IgE will be aggregated on exposure to allergen.

1-9-3-8 Natural killer cells

Natural killer (NK) cells are an element of the innate immune system that attack infected cells, but do not directly attack invading microbes. They attack and destroy tumor cells and virus infected cells through a process called missing-self.

1-9-4 Factors affecting the immune system

1-9-4-1 Anatomic Barriers:

a. Skin - low pH due to lactic and fatty acid
b. Epidermis - epidermal cells are water proofing
c. Dermis - very low pH (3-5) due to lactic and fatty cells, under the skin layer, and has sebaceous glands with hair follicles
d. Mucous membranes - a huge surface area that include saliva, tears, ciliated epithelial cells in respiratory tracts, GI (gastrointestinal), and Urogenital

1-9-4-2 Physiological barriers:

a. pH - low pH helps inhibit microbial growth
b. Temperature and pressure - normal temperature and pressure (37° C and 1 bar)
c. Oxygen, carbon dioxide, and pollutants
d. Other adaptive barriers include pepsin (digestive enzyme that hydrolyzes proteins), lysozymes (enzyme able to cleave cell wall of bacteria), cryptidine and defensine (found in the base of the small intestine, complements, and interferon (protein produced by the body upon infection, and inhibit viral replication)

The question is "can microorganisms resist our immune system?" The answer is yes.

First let's talk about the complement pathways in detail, and then we shall see the mechanism of resistance of microorganisms to the immune system.

In the compliment pathways, there are over 20 complexes of proteins which are produced in the liver and circulate in the plasma. These proteins (enzymes) will be activated molecules when they are in contact with microorganism and other antigen and antibody molecules. The complement system has 9 components (C1 - C9) scattered in the plasma. When activated by microorganisms, each component will split into two fragments, one large and one small. For example, activation of the C4 will have fragments of C4a and C4b. The designation "a" stands for the larger fragment and "b" for the smaller. In addition to components C1 to C9, there are other regulating components that regulate the activation of C1 to C9. For example, properdin P, factors B and D, C3 convertase, etc, Figure (53).

Figure (53): Components of complement system

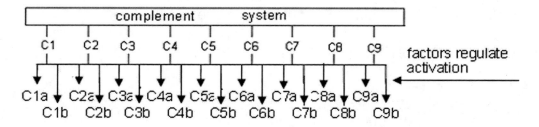

The functions of the above fragments are as follows:

- C3a, C4a, and C5a will activate the mast cell which will release vasodilators to widen the blood vessels. This process is called anaphylotoxins.
- The fragment C3b binds to the surface of some bacteria and then engulfs them. The white blood cells and macrophages will kill the engulfed bacteria. The coating of bacteria is called opsonization, and the engulfment is called phagocytosis.
- The fragments C3a and C5a attract white blood cells and microphage to the site of infection through a process called chemotaxis.
- Other fragments and components such as C5b, C6, C7, C8, and C9 combine on the surface of some microorganisms and form pore-like structures called Membrane Attack Complex (MAC's), or membrane lesions that will make holes in the microorganisms leading to cell rupture and death, Figure (54).

Figure (54): Components and fragments of complement system

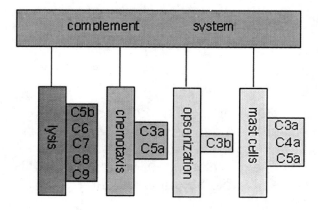

Some microorganisms have created a defense against opsonization by surrounding themselves with a coat and forming a capsule called macoplysaccaride. In some cases the receptors of our immunity don't fit the pores of evolved microorganism, Figure (55).

Figure (55): Bacterial cover preventing C3b on phagocytes from bonding to C3b attached to a bacterium cell wall

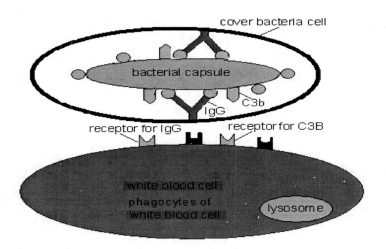

One of the functions of certain microorganism molecules known as IgG is to stick antigens such as protein and polysaccharides to phagocytes. The IgG and C3b of antigens can bind to fit the receptor of phagocytes. However, antigens can evolve and build a new cover onto their walls which prevents binding with phagocytes. The process of binding is called opsonization.

1-9-4-3 Other factors

- o Age - The immune system may be enhanced with age because one acquires stronger and more versatile immunity to more diseases as viruses evolve into different shapes and strengths. However, some immune molecules become weaker with age, and their responses are more sluggish and less effective to invading microorganisms.
- o Allergies - pollen and dust are harmless to humans but they overreact with the immune system, and they trigger the release of IgE (the antibody immunglubin E) that in turn releases inflammatory particles such as histamine. The histamine helps lead to the symptoms of allergies such as hay fever, rhinitis, hives, and asthma.
- o Heredity - is the traits and characteristics of your genes and immune system that are composed from proteins, lipids, and sugar. The structure of immune molecules is dependant on genes. Genetic structure can play important roles in determining the strength of the immune system to attack different species. For example, your wife can withstand the flu coming to your home, but you cannot.
- o Autoimmune disease - antibodies created by your immune system to attack your own cells, as if your cells are seen by your immune system as pathogens and antibodies. For example, liver diseases such as cirrhosis that are caused without hepatitis C or B, or without much alcohol drinking, could be related to the autoimmune disease.

Autoimmune disease is associated with many disorders such as chronic muscle fatigues, pancreatic disorder (diabetes), etc. Researchers don't find causes of such self-destructive autoimmune disease. Some types of cancers and rheumatoid arthritis are attributed to autoimmune disease. Researchers are using gene therapy to weaken the autoimmune molecules and strengthen the immune system by injecting modified genes into tumor cells and other location of the body, hoping that the gene therapy will kill the tumor cells. The mystery in the battle between our immune system and the immune system of pathogens and antibodies is that both immune systems are evolved to compromise each other.

1-9-5 Adaptive immunity

Innate immunity, by itself, may not be sufficient to protect our body against pathogens or invading microorganisms. However, if innate immunity fails to protect the body, then the adaptive immunity will intervene. Both the innate and adaptive immune systems have different ways to react:

a) The innate immune system reacts immediately, where the adaptive immune takes some time to react.
b) The innate immune system is general, i.e. is not specific in attacking pathogens, and reacts equally to all types of invaders, whereas the adaptive immune system can distinguish between invaders, which makes it antigen-specific.
c) The innate immune system does not have the ability to remember types of previous invaders. The adaptive immune system reacts more rapidly towards recognized antigens and pathogens because it memorizes their types. It is highly adaptable, due to the hypermutation in the somatic cell (all cells except the sperm and egg cells), and due to the recombination of V(D)J cells (adjustment of shape and location of T and IgG cells to match the invaders' cells).

The cells of the adaptive immune system are a type of leukocyte, called a lymphocyte. Lymphocytes are small white blood cells that bear the major responsibility for carrying out the activities of the immune system; they number about two trillion. Lymphocytes are one of the five kinds of white blood cells or leukocytes, circulating in the blood. The two major cells of the lymphocytes are B cells and T cells. In nearly all vertebrates, B and T cells are produced by stem cells in the bone marrow. T cells travel to and develop in the thymus gland. Both B and T cells recognize specific antigen targets. B cells complete their maturation in the bone marrow. Inside the thymus gland. T cells educate themselves to distinguish self cells from non-self cells. Figure (56) shows the organs and nodes of the immune system.

Figure (56): Lymphatic nodes

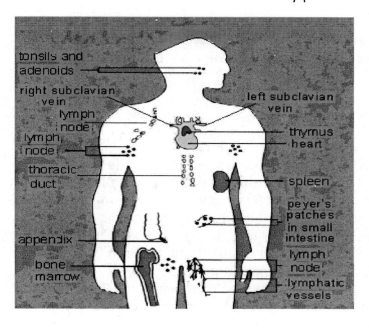

1-9-5-1 Stages of B and T cells

- Naïve cells have matured and left the bone marrow or thymus, and have entered the lymphatic system to become ready for attacking their cognate antigens,
- Effector cells are currently involved in the attack,
- Memory cells that are the longest to live past infection.

The bone marrow produces 'hematopoietic stem cells' which produce the main two types of blood cells, Figure (57).

Figure (57): Immune cells produced from hematopoietic cells.

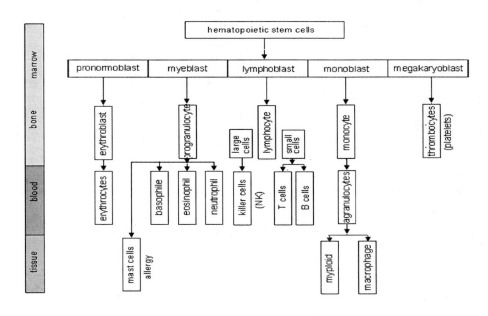

Myeloid cells including monocytes, macrophage, neutrophils, basophiles, eosinophils, eurothocytes, platelets and dendretic cells,
  o   Lymphoid cells including T cells, B cells, and Nk killer cells

1-9-5-2 B cells

Each B cell is programmed to tackle one specific pathogen. For example, one B cell is programmed to block a virus that causes pneumonia, and another B cell blocks a virus that causes allergies. When B cells activate, they initiate the plasma to produce millions of identical antibody molecules to encounter the triggering antigens. A given antibody matches an antigen like a key matches a lock. The fit varies, sometimes it is very precise, while at other times it is little better than that of a skeleton key. To some degree, however, the antibody interlocks with the antigen and thereby marks it for destruction. Each B cell is programmed to make one specific antibody. For example, one B cell will make an antibody that blocks a virus that causes the common cold, while another produces an antibody that zeros in on a bacterium that causes pneumonia, http://www.immunecentral.com/immune-system/iss7.cfm.

The B cell and interaction with plasma cells is shown in Figure (58).

Figure (58): Produced antibodies in the plasma cells due to the activation of B cells.

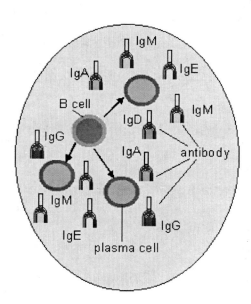

There are nine kinds of Ig antibodies (immunoglobins) produced by plasma cells: 4 of IgG, 2 of IgA, and I of each of IgM, IgE, and IgD. Each kind has a distinct role in the way they attack foreign bodies:

  o   IgG provides coating

- IgA blocks microorganisms from entering gastrointestinal tracts, saliva, and fluid tears
- IgM combines in star-shaped clusters to be readily available in the blood stream
- IgE defends against allergies and parasites
- IgD regulate the activation of b cells

1-9-5-3 T cells and Lymphokines

When T cells mature, they will be divided into two types: regulatory T cells, and cytotoxic cells, Figure (59).

Figure (59): T cell's components

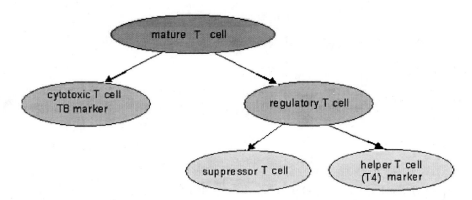

Helper T cells are essential for activating cells, other T cells, macrophages and killer cells of the innate immune system. Cytotoxic T cells are killer cells, and they get rid of the cells that have been infected by virus or cancer. In addition, they are responsible for the rejection of tissue and organ grafts.

The T4 and T8 cells determine the type of class either class I or class II of MHC (major histocompatibility complex). Figure (60) shows the interaction between T8 and T4 cells with virus, macrophage, and B cells.

Figure (60): Treatment of T cells to antibodies, and self cells

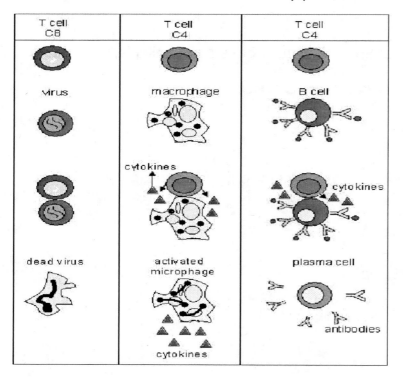

The T cell has two components; the receptor (T-cell receptor; TCR), and the CD (CD4 and CD8), which stands for cluster of differentiation. CDs indicate a defined subset of cellular receptors (epitpes) that identify cell type and stage of differentiation (progress), which are recognized by antibodies. There are about 250 CDs of different molecules on their surface, coating B cells and T cells. CD4 and CD8 are the most active molecules that respond to the movement of T cells towards macrophages and B cells.

The interaction between T cells and the antigenic peptide which is bonded to histocompatibility (MHC) is the core of the immune system. The infected cells are marked with a combination of the foreign protein and the MHC molecule. The combination links to a specific TCR. The goal of the link is to destroy the damaged (infected) cell. New research indicates that the invading virus produce something called 'antigen peptide' within the infected cell. This antigen peptide is held within the MHC like a hotdog in a bun. The bun is then engulfed by the TCR. There are two types of MHCs: MHC class I which binds to CD8 and MHC class II which binds to CD4. The process of killing the invading virus becomes simple to the TCR:

- virus invades a healthy cell
- the infected cell produces fragments of the viral protein (peptide antigen) on its surface

- o the peptide antigen binds to a molecule of MHC class I or MHC class II depending on the type of the viral protein whether it is one glycoprotein (MHC class II) or two glycoproteins (MHC class I)
- o the T cells (T4 or T8) monitor the class of MHC, and then the cell binds to it and destroy it before it can release a fresh crop of viruses.

Above steps are shown in Figure (61).

Figure (61): Interaction between T cell and MHC

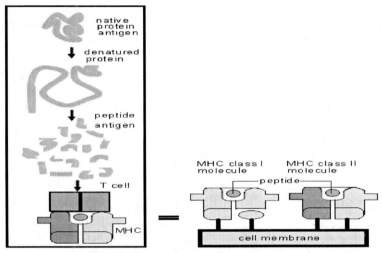

Redrawn from http://people.rit.edu/~gtfsbi/imm/20082part1.htm

When T4 cells bind to macrophages or B cells, they produce activators called 'cytokines' which are critical to the development and functioning of both the innate and adaptive immune response, although they are not limited to just the immune system. In addition to their activation and recruitment of immune cells to increase the response to the invading microorganisms, they play important roles during embryogenesis.
Figure (62) shows the TCR and the CD4 binding to a complimentary MHC class II molecule with attached peptide epitope on an activated B cell. The ctitokines enable the activated B cell to proliferate, differentiating the b cell and plasma cell.

Figure (62): Interaction between TCR and B cell

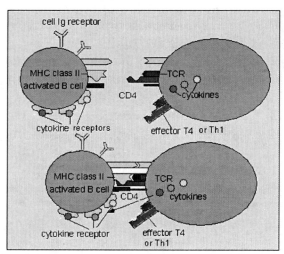

Redrawn from
http://student.ccbcmd.edu/courses/bio141/lecguide/unit5/intro/bcell/bcellt4_fl.html

The immune system protects against four classes of pathogen, Table (7).

Table (7): types of pathogens

Type of pathogen		Examples	Disesase
Bacteria		Salmonella enteridis	Food poisoning
		Myrobacterium tuberculosis	Tuberculosis
viruses		Variola	Small box
		Influenza	Flu
		HIV	AIDS
Fungi		Epidermopyton floccosum	Ringworm
		Candiad albicans	Thrush, systemic candidiasis
parasite	Protozoa	Trypanosome brucel	Sleeping sickness
		Leishmania donovari	Leishmaiasis
		Plasomodium falciparum	Malaria
	worms	Ascaris lumbricoides	ascariasis
		Schistosoma mansoni	schistosmiasis

1-9-6 Structures of the immune system

The immune system is organized and structured in two lymphoid tissues; the central (bone marrow and thymus), and the peripheral (lymph nodes, spleen, mucosa-associated lymphoid tissue).

1-9-6-1 Bone marrow

Red blood cells, white blood cells (lymphocytes and macrophages) and platelets are manufactured in the bone marrow. All the cells of the immune system are coming from stem cells in the bone marrow.

77

2-9-6-2 Thymus gland

Lymphoid cells mature and educate themselves before they are released into the circulation. When they are released, the T cells are self tolerated and adjusted to perform effective destruction to pathogens.

T cells enter the cortex proliferate, mature and pass to the medulla from the medulla they pass to the circulation system, Figure (63).

Figure (63): components of thymus gland

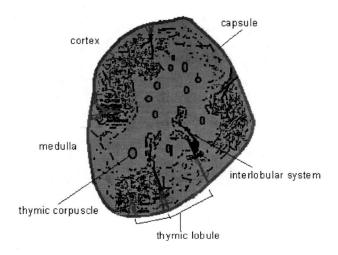

The volume of the thymus gland gradually increases from childhood to puberty and then stops at an age when functioning starts to slow, but it still functions.

1-9-6-3 lymph nodes

Lymph nodes are amassed in some locations such as the neck and groins. Axillae, and para-aortic area. The main functions of the lymph node are:

The lymph node has the phagocytic cells which filter microorganisms and pathogens,

The lymph nodes present antigens to the immune system.

The node has several components: sinuses, parenchyma (medulla, cortex, and paracortex), follicles, germinal centres, and lympahitcs, Figure (64).

Figure (64): components of the lymph node

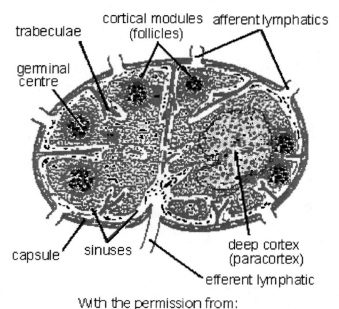

trabeculae

germinal centre

cortical modules (follicles)

afferent lymphatics

capsule   sinuses

deep cortex (paracortex)

efferent lymphatic

With the permission from:
http://www.microbiologybytes.com/iandi/2b.html

The lymph enters the node through the afferent lymphatic to the medulla deep inside, and microorganisms are removed by macrophages.

1-9-6-4 Spleen

The spleen is located in the upper left quadrant of the abdomen. The spleen functions in the destruction of redundant red blood cells, and holds the blood in its reservoir. It is a part of the immune system, and its malfunction leads to tendency to certain infections.
It has two main areas:

- The red pulp area which filters red blood cells,
- The white pulp which stores B and T lymphocytes (cells).

The red pulp consists of large numbers of sinuses and sinusoids filled with blood and are in charge of filtration of the blood. The white pulp consists of lymphoid tissue and is in charge for the immunological function of the spleen, Figure (65).

Figure (65): The spleen with its components

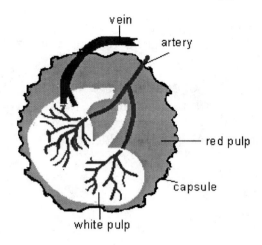

1-9-7 Circulation in the lymph system

Activated T cells present in the lymph nodes could divide, produce effector cytokines, and migrate to peripheral tissues. However, activated T cells that migrated into nonlymphoid tissues (airways and lung tissues) produce substantial numbers of effector cytokines, but cannot divide or migrate back to the lymph nodes. So, activated T cells, upon antigen challenge, can undergo clonal expansion in the lymph nodes. But they are recruited and retained as non-dividing cells in nonlymphoid tissues. T cells can accumulate in large quantities inside nonlymphoid tissues and can diffuse and cause extranodal lymphoma. They could be more aggressive than lymph nodal lymphomas that occurr in stomach, tonsils, skin, small intestines, breasts, and salivary glands. Figure (66) is of the circulation of the immune molecules, and for the lymph nodes and nonlymphoid tissues.

Figure (66): Circulation of lymphatic System

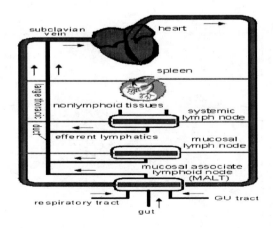

1-9-8 Mechanism of the Immune System

One of the important mechanisms in fighting pathogens is accomplished in three steps:

1. the antibody directs itself towards the target cell (invaded cell), and then triggers killer cells to kill or block the action of cell survivor factors, such as growth factors
2. Igs cells (particularly the Fc portion), Figure (67) will bind to the MHC of the antibody.

Figure (67): Ig molecule with Fc and Fab branches

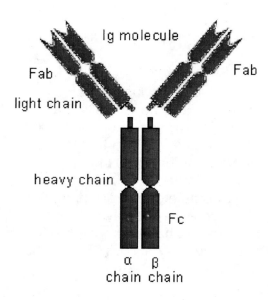

3. The Fab will bind to the receptor of the invading microorganism, and release something called FcgRs, and kill the invader, Figure (68).

Figure (68): Ig molecules blocking bacteria

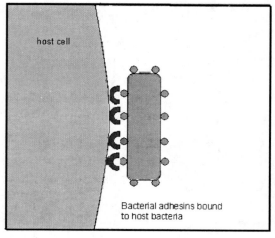
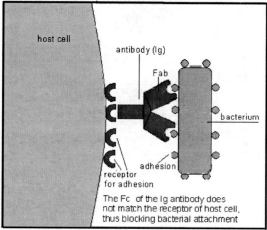

B cells have surface receptors like T cells, but do not interact directly to neutralize or destroy the antigen. When they recognize the antigen, they produce daughter B cells of the same shapes and functions as themselves. The new B cells develop into short-lived plasma cells that produce antibodies and release them into the circulation system of the lymph nodes. Not all B cells produce daughters, some of them reside in the main lymphoid tissues and turn into memory cells which continue to produce small amounts of antibodies long after an infection has been destroyed, Figure (69).

Figure (69): Interaction between B cells and antigens

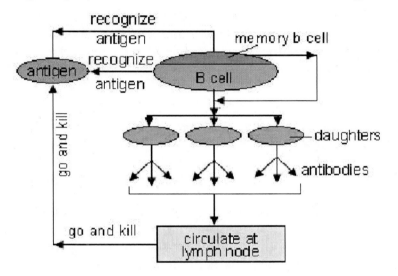

Millions of B cells, each with different binding mechanisms are produced in the blood plasma. Daughters are forms of clones of identical cells. The circulation of the antibodies is similar to T cells because B cells have similar gamma globulins (Igs) that combine with antigens and destroy them. From Figure (68), we can see B cells also divide and produce new clones of B cells.

Gamma globulins (Igs) are now being used to reduce antibodies in patients with organs transplantation and organs failure to reduce antibodies that are produced by the disease.

1-9-9 Disorders of the immune system

Disorders of the immune system can be broken down into four main types:

Immunodeficiency disorders (primary and acquired)
Immune system disorders occur when the immune system does not fight invading microorganisms, tumors or harmful substances as it should. The immune system may be hyper active or hypoactive. When lymphocytes T and B

or both do not function properly, then the immune system has a in deficiency disorders, and does not produce enough antibodies.

The primary immunodeficiencies (inherited) that affects B cells includes:

- hypogammagloblinemia which affects the respiratory system, causing infections,
- a gammaglobulinemia which results in severe infections and is often deadly.

The acquired immunodeficiency can be developed due to non friendly environmental incidences such as disease, burns, drugs, pollutions, malnutrition, etc. The acquired immunodeficiency includes

Immunodeficiency caused by medicinal drugs. For example, medicines given to patients with transplant organs that suppress the immunity, could cause immunodeficiency. Radiation and chemotherapy treatment for cancer attacks in addition to cancerous cells, healthy cells.

HIV (human immunodeficiency virus) which is an acquired immunodeficiency syndrome, it is slowly and steadily destroys the immune system which wipes out T-helper cells. HIV can be transferred to newborns from their mothers while in the uterus or during breast feeding. HIV patients can get the disease through unprotected sexual intercourse, infected needles, tattoos, or steroids.

Autoimmune diseases are disorders in which the body's immune system reacts against some of its own tissue and organs, and produces antibodies to attack itself:

Juvenile rheumatoid arthritis is a disease in which the immune system sees certain parts of the body such as the knees, hands, feet, elbows, etc as foreign organs and attacks them.

Lupus is a disease attacking muscles and includes joint inflammation and pain, and also may attack the kidneys. Lupus can attack the blood cells and reduce the number of red or white blood cells or platelets.

Juvenile dermatomyositis is a disorder noticeable by inflammation and damage of the skin and muscles. It causes vasculitis that manifests itself in children. It is the pediatric counterpart of dermatomyositis.

Scleroderma is a chronic disease that causes damage and inflammation in the skin, joint, and internal organs. Scleroderma causes formation of scar tissue (fibrosis) in the skin and organs of the body. This leads to thickness and firmness of involved areas and is called 'systemic sclerosis'.

Allergic Disorders occur when the immune system overreacts to exposure to antigens in the environment. The microorganisms that cause such attacks are called allergens. The immune response can cause symptoms such as swelling, watery eyes, sneezing, and even a life-threatening reaction called anaphylaxis. Taking antihistamine medications can relieve the symptoms. Allergic disorders include:

    a. Eczema causes rash known as atopic dermatitis, infantile, and flexural, in kids and teens that have allergies such as hives, hay fever, or asthma. It is known as infantile e., flexural e., atopic dermatitis.

    b. Asthma causes breathing difficulties due to an allergic response in the lungs. Allergens, such as dust mites, molds, pollen, animal dander and certain fibers, can trigger breathing tubes in the lungs and reduce their diameters, leading to difficulty in breathing.

    c. Different types of allergies, such as environmental, seasonal, drug, food, and toxins can cause different disease from the flu to hives.

1-9-10 Cancers of the immune system, see Chapter (3)

Cells of the immune system and other body systems can proliferate uncontrollably resulting in cancer. The cancer can weaken the immune system if it spreads in the bone marrow where T and B cells are made. This happens most often in leukemia and lymphoma. The cancer in the bone marrow stops the bone marrow from making so many blood cells; the white, the red, and the platelet cells. Leukemia is caused by the proliferation of white blood cells (leukocytes) uncontrollably. The uncontrolled growth of antibody producing cells can lead to many types of cancer such as myeloma, lymphoma, melanoma, non-Hodking lymphoma, AIDS-related cancer, and many other types of cancer. Many types of cancer can be treated very successfully by drugs, chemotherapy, and/or irradiation. Chapter 3 concentrates on cancer.

1-9-10-1 Vaccines

A vaccine is any preparation by medical drug companies, intended to produce immunity to a disease by stimulating and boosting the production of antibodies. Vaccines include delays and suspension of destroyed or attenuated pathogens and antigens, or products or derivatives of microorganisms. The most common method of administering vaccines is through inoculation, by mouth, by use of nasal spray, or by an intravenous drip. Vaccines may be inactivated or dead organisms or modified products derived from them. There are four different types of traditional vaccines:

o Killed microorganisms vaccine: Microorganisms are killed by chemicals or heat and used against flu, cholera, some types of plague, and hepatitis A in the liver. In the preparation of such vaccines, two flu strains are combined together, and a new flu is produced for vaccination, Figure (70).

Figure (70): HA and NA of two different types of flu are combined together in such a way that two of the first strain and six of the second are cultivated in a special plasmid to produce a new flu strain.

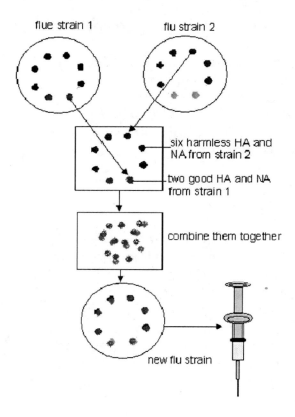

- Attenuated living microorganism: Live microorganisms with disabled characteristics and without dangerous effect are cultivated in certain environments and used against diseases such as measles, rubella, fever, mumps.
- Fragments of antigens (note that antigens are proteins residing on the surface of pathogens): This type of vaccine uses attenuated antigens. The vaccine can be used against hepatitis B (hepatitis A, B, and C are related to the liver and the worst is hepatitis C).
- Toxoids against toxins: Toxoids are used against snake bites, and scorpion and black widow stings. Toxoids are prepared through chemical processes in which inactivated toxic chemicals are moderated by some catalysts.

1-9-10-2 Risk of vaccines

Vaccines are proteins that are cultured in animal tissues such as horse blood, monkey's kidneys, rabbit, and even human fetal lung tissues. Proteins cultured in animals may contain animal viruses, and could cause disorders in the human body. Since it is impossible to isolate unwanted viruses from others within the animal tissues, some vaccines contain risky viruses that could deteriorate the situation of patients. For example, the AIDS virus was spread from monkeys to humans through contaminated polio vaccines given to Africans. Some new innovative vaccines (DNA combined to protein antigens) causes non-Hodgkin's lymphoma in Baby Boomers.

Some dangerous chemicals are being used as preservatives and they are adjuvant to vaccines. Such chemicals are dangerous; they include aluminum phosphate, formaldehyde, phenol (hexagonal carbolic acid) acetone, ethyl mercury, and others. Flu vaccines still contain mercury. These additives could lead to many health disorders such as Alzheimers, blindness, autism, liver and kidney failure, and cancer. However, there is no certain evidence linking these disorders with vaccines.

Germs are not necessarily dangerous. The "vaccine theory" is based upon the "germ theory", which says that disease is caused by germs. The germ theory itself has always been in question. We encounter billions of germs everyday but we do not get sick everyday. The "terrain theory" hypothesizes that viral infections are not caused by germs, but by toxic conditions in the body, which make it a favorable host for these organisms. Germs can have a mutually beneficial relationship with our bodies by feeding on and helping eliminating toxins. So, the key to combating infectious disease is not to kill the germ but to make the body an unfavorable host. Feed it healthy sources of energy and don't poison it with toxins. Vaccinations introduce toxins into the body and therefore actually cause illness,
http://www.vran.org/docs/VRAN-Immunization-Fact-Sheet-v6.pdf.

There is growing evidence that immunization can cause a large number of other chronic diseases including diabetes, obesity, metabolic syndrome, autoimmune diseases, allergies, asthma, cancers, and Gulf War Syndrome. Data linking these diseases to vaccines includes human and animal data. In some cases the increased risk of developing these diseases following immunization exceeds the risk of the infectious complications prevented by immunization. The content of this site is not intended to be anti-immunization but instead to promote the concept that the goal of immunization is to promote health not eradicate infections, http://www.vaccines.net/newpage114.htm.

1-9-11 Transplantation immunity

Regeneration of cells, including T cells and humoral cells (B cells), may offer a promising advance in transplantation of organs such as the liver, kidneys, heart and lungs. The lymphocyte cells (T and B) recognize the foreign antigens of transplanted tissues and kill them. This action would lead to organ failure, if not treated successfully. Peripheral T cells of CD4+CD25+ regulate the transplantation allograft tolerance, and their counts vary according to the degree of graft acceptance. Certain drugs can modulate the number of count; otherwise if the count increases the transplanted organ may fail. Some drugs such as tacrolimus, cellcepts (MMF), steroid, asathioprine and other immunosuppressant proved to be effective in keeping the count acceptable.

After a transplant, the immune system is reconstituted if the derived hematopoietic cells of the organ recipient matches and combines with the peripheral T cells from the donor. The antigen specific donor T cells function will be poor if the antigen cells from the hematopoietic cells encounter delay or deficiency. Rejection to the new organ may be classified into three types:

- o Rejection due to unmatched blood types. There are four blood types : O, A, B, and AB, Figure (71). Before the transplant operation, it is important to correctly determine a patient's blood type, or prior to transfusion of any blood component. It is also important to determine the RhD (rhesus D/protein carrying blood group D) antigen which is the negativity or positivity of the blood type. For example, blood type O can be O negative, or O positive. Unmatched blood type could lead to immediate death.

Figure (71): Blood types

type red blood cell	A	B	AB	O
antibodies	anti-B	anti-A	none	anti A&anti B
antigens	A-antigen	B-antigen	A&B antigen	none

- o Rejection due to mismatch of the leukocyte HLA on MHC class I and class II (demonstrated earlier). This type of rejection can happen in the first three months after transplantation. However, it can also occur months to years after transplantation.
- o Chronic rejection is a long loss of function in the transplanted organ due to fibrosis or scars formed in the organ. It can also be caused by the a part of the MHC, but not by the HLA, as in the above type of rejection such as the H-Y gene of the male chromosome. This type can be treated by adjusted doses of corticosteroids.

1-10 Human respiratory system

Air enters the nostrils to the lungs until it reaches the alveoli, Figure (72).

Figure (72): Components of the lung

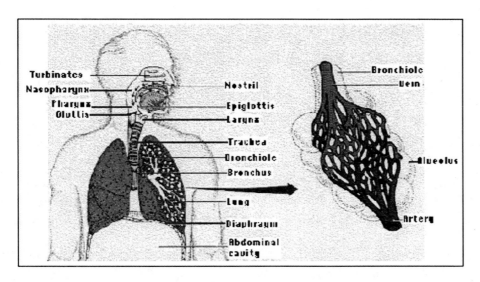

http://users.rcn.com/jkimball.ma.ultranet/BiologyPages/P/Pulmonary.html

The gas exchange takes place in the alveoli. The two lungs have about 300 million alveoli that could provide an area of about 150 square meters.

Our body constantly needs energy that is produced when sugar and fats are combined with oxygen and burnt. The oxygen is taken through the lung into the red blood cells. When the air is introduced into the thoracic cavity (the diaphragm divides the body cavity into the thoracic cavity which contains the lungs and heart, and the abdominal cavity which contains the stomach and intestines), the volume increases and causes the lungs to expand. The respiration process has two mechanisms:

o   Inspiration (inhaling) in which the intercostal muscles contract, lifting the ribs up and out. The diaphragm contracts, drawing it down.
o   Expiration (exhaling) in which the above mechanism is reversed.

At rest, we breathe 15 to 20 times a minute exchanging about 600 ml of air. This depends on the body weight and the surrounding environment. In a larger environment, we breathe about 850 liters of air. Table (8) shows the percentage of air coming in and out of the lungs.

Table (8): In and Out air of the lung

Air component	air in %	Air out %
N_2	78.62	74.9
O_2	20.85	15.3
CO_2	0.03	3.6
H_2O	0.5	6.2
Total	100	100

1-10-1 Diseases of the lungs

- Asthma is a constriction of bronchi and bronchioles which makes it more difficult to breathe in and out. Asthma can be triggered by airborne irritants such as cigarette smoke and chemical fumes, and by some types of food (peanuts).
- Pneumonia is an infection of the alveoli that is caused by certain bacteria and viruses. Supplemental oxygen may be needed if the affected area of alveoli is large.
- Chronic bronchitis is an increase in the amount of mucus secreted in the bronchi and bronchioles. The air passage becomes clogged with mucus, and it leads to a persistent cough and noise in breathing.
- Emphysema is the worst among all lugs diseases. It is associated with the break down of the alveoli's walls. This disease could lead to death due to the loss of exchange area between oxygen and carbon dioxide that forces the heart to pump a larger quantity of blood to compensate the loss in gas exchange. Emphysema is thought to be genetic due to the defective genes alpha-1-antitrypsin. Luckily, emphysema can be treated by the same gene inhibitor.
  Some patients can have more than one of the above disease in the same time.
- Lung cancer is the most common cause of death in the U.S. for males. Recently, women have been dying in larger numbers from lung cancer than breast cancer. Lung cancer is caused when the epithelial cells lining the bronchi and bronchioles proliferate uncontrollably. Metastasis from other parts of the body can move to the lungs and cause cancer. Tumor metastases are made up of the same type of cells as the original. For example, if colon cancer spreads via the blood stream or the lymph system to the lungs, it is metastatic colon cancer in the lung, not lung cancer.

Figure (73) shows a lung cancer.

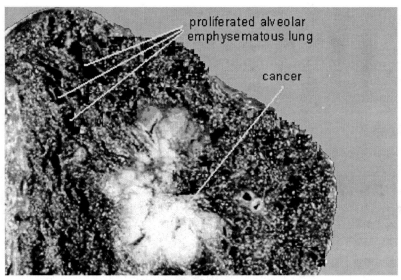

redrawn from
http://www.medicinenet.com/lung_cancer/article.htm

1-11 Hormones

In 1989, at Veterans Administration hospitals in Milwaukee and Chicago, a small group of men aged 60 and over began receiving bio-identical hormone injections three times a week that dramatically reversed some signs of aging. The injections increased their lean body (and presumably muscle) mass, reduced excess fat, and thickened skin. When the injections stopped, the men's new strength ebbed and signs of aging returned. Human growth hormone levels decrease in about half of all adults with the passage of time. Production of the sex hormones estrogen and testosterone tends to fall off. Hormones with less familiar names, like melatonin and thymosin, are also not as abundant in older as in younger adults.

Hormones are chemicals produced in multicellular organs that interact with receptors in target tissues to change the function of that tissue. There are mainly two types of hormones produced; the endocrine and the exocrine hormones. The endocrine system is a set of glands that secrete chemical peptides of hormones. These hormones are passed through the blood to arrive at a target organ, which has cells possessing the appropriate receptor. Exocrine glands secrete enzymes (not hormones) that are passed outside the body. Sweat glands, salivary glands, and digestive glands are examples of exocrine glands. The pancreas produces digestive enzymes and is called an exocrine gland. At the same time it produces Insulin, which is a hormone, and is also an endocrine gland.

Hormones control and regulate a number of essential functions in the body. For example estrogen in women controls pregnancy and emotion. Insulin and glucagon is excreted in the pancreas to control sugar levels during ingestion and hunger. Thyroxin is produced in the thyroid gland to regulate the metabolic rate.

1-11-1 Endocrine system

The endocrine system is one of main system in the body for controlling, coordinating and communicating the body's function. It interacts with all body's systems such as the liver, kidney, heart, pancreas, and reproductive system to control and maintain:

- Body energy level
- Growth and development
- Homeostasis (temperature, pressure, pH, chemical, etc)
- Neurological balance and senses
- Reproduction

The endocrine system achieves the above functions through a distribution and supply system consisting of glands and organs that produce, secrete and store certain types of hormones. Hormones are chemicals that react with cells and organs' chemicals (receptors) to help them in restoring, correcting, or modifying functions, see Figure (74) for the endocrine system.

Figure (74): Major endocrine glands in the body

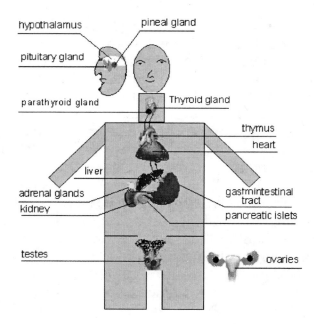

Hormones produced by the endocrine system are listed in table (9).

Table (9): Hormones of the endocrine system

Endocrine gland	Hormone	Effect

Hypothalamus	o Corticotrophin	Neurotransmitter polypeptide hormone released when there is a stress.
	o Somatostatin	Regulate the endocrine system.
	o Dopamine	It is a neurotransmitter activating five of the brain transmitters. It is produced in the substantia nigra (in brain) and the ventral segmental area. It blocks the release of prolactin (controls lactation of breast feeding) from the interior lobe of the pituitary.
	o Thyroprotein	It stimulates the release of thyroid and prolactin hormone.
Pituitary gland is about the size of a pea, protruded off the bottom of the hypothalamus at the base of the brain.	o ACTH o TSH o PRI o GH o LH o FSH o Endorphin	All of them are responsible for regulation and control of growth, blood pressure, sex organs, breast milk, thyroid gland, conversion of sugar to ATP, and liquidity.
Heart	o Atriopeptin	It is a hormone that controls blood volume by increasing the rate of urine when the blood volume is increase.
Adrenal gland	The cortex:	
	o Cortisol	Cortisol (glucocorticoid) stimulates the glucagon to convert fat and protein into glucose during fasting or hunger. The hormone can also reduce inflammation in the body, and therefore it can be used in therapy of rheumatoid arthritis, against autoimmune diseases, and to prevent the rejection of transplanted organs. However, cortisol can help increase the weight of patients with transplant

92

livers due to the increased fat in the new liver.

- o Aldosterone

It causes the tubules of the kidney to retain water and sodium, thus increasing the water and the level of salt in the body, and it consequently increases the blood pressure. Blocking the aldosterone receptor can reduce the blood pressure. The aldosterone hormone regulates the blood pressure.

- o Androgen (Testosterone)

The medulla:

Androgen promotes masculine characteristics and affects hair growth.

- o Epinephrine (adrenalin

- o Norepinephrine

It is secreted by the adrenal medulla in response to stress and stimulates autonomic nerve action.

Kidney

- o Erythropoietin

A hormone as adrenalin and a neurotransmitter. Erythropoietin (EPO) is a glycoprotein that simulates the bone marrow to produce more red blood cells. Bleeding or moving to higher altitude with little oxygen stimulates the EPO. Patient with failed kidney do not produce enough EPO, and they suffer from anemia.

- o Calciferol

Is converted in the liver to vitamin B3 which then goes to the kidney to be converted back to calciferol. The final step is stimulated by the parathyroid hormone PTH

| | o Rennin | Rennin is an enzyme to regulate the blood pressure: as shown in Figure (75) below. |

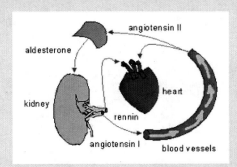

Rennin enters blood vessels as angiotensin I, which is converted to angiotensin II by the peptidase produced by blood vessels. Angiotensin II constricts the walls of the arterioles, narrowing capillaries in the heart, and stimulates the kidney's tubules to absorb sodium and retain water. When angiotensin II passes through the adrenal cortex (the adrenal gland has cortex and medulla), it releases aldosterone. Aldosterone prompts the kidney to reclaim more sodium and water. The accumulation of sodium and water in the kidney causes the heart to beat faster, which calls the posterior of the pituitary gland to release the vasopressin hormone that reduces the urine flow.

Is converted in the liver to vitamin B3 which is then goes to the kidney to be converted back to calciferol. The final step is stimulated by the parathyroid hormone PTH

Pineal gland   o Melatonin   Melatonin is important in the regulation of the circadian rhythms of several biological functions. Circadian cycle is constant unless there is a cue change such as light and temperature

Thyroid gland	o Calcitonin	It regulates calcium homeostasis.
	o Thyroid T3 and T4	Both regulate the rate of metabolism and the rate of grow.
Thymus gland It stimulates the production of white blood cells. After puberty, the gland begins to shrink and is replaced by connective tissue and fat. The main function of the thymus gland is in the processing and maturation of special lymphocytes called T-cells	o Tymosins	Thymosins help the maturation of lymphocytes in the gland, as well as the growth and activity of lymphocytes.
	o Thymopoietin	Thymopoietin is a gene which encodes three mRNAs encoding proteins. It helps the production of CD90.
Gastrointestinal tract	o Gastrin	A hormone released after eating that control the amount of acid in the stomach. It is secreted by the mucous lining of the stomach.
	o Colecystokinin	Gastrointestinal hormone that stimulates the secretion of pancreatic enzymes and the contraction and emptying of the gall bladder for more bilerubin to digest food, particularly the fat.
	o Secretin	It is a peptide hormone produced in the duodenum. Its primary effect is to regulate the pH.
	o Motelin	Motilin is an intestinal peptide that stimulates contraction of the guts smooth muscle.

	o Ghrelin	A peptide hormone secreted by cells in the lining of the stomach. Ghrelin is important in appetite regulation and maintaining the body's energy balance. Extra amounts of secreted ghrelin leads to obesity.
	o Enteroglucagon & Glucagon like-peptide	Enteroglucagon is released following ingestion of a mixed meal, and delays gastric emptying. It is released in the colon and terminal ileum.
	o Gastric inhibitory peptide (GIP)	It is believed that the function of GIP is to stimulate insulin secretion.
	o Vasoactive intestinal peptide	It induces smooth muscle relaxation, and stimulates secretion of water into pancreatic juice and bile.
Pancreatic islets	o Glucagon	A hormone produced by the islets of the Langerhans cell of the pancreas, to increase the level of sugar in the blood, thus opposing the action of insulin.
	o Insulin	A hormone produced by the endocrine islets of Langerhans cells of the pancreas. It acts to burn the sugar (oxidizing) and thus lower blood sugar levels.
	o Somatostatin	Somatostatin helps tell the body when to make the other hormones, such as insulin, glucagon, gastrin, secretin, and rennin. It helps block other hormones.
	o Pancreatic polypeptide	It is released after food intake and it is believed to act as a regulator of pancreatic and gastrointestinal functions.

96

Testes (male)	o Androgen	It is a hormone that produces the sexual physical characteristics of men (such as deep voice and facial hair). The main male androgen is testosterone.
	o Inhibin	A hormone produced by the testes and ovaries which regulates FSH (controls ovulation in females and sperm maturation in men) levels, and thus, spermatogenesis.
Ovaries (female)	o Estrogens	A hormone produced in the ovaries, which is involved in the menstrual cycle and in developing and maintaining female sexual characteristics (e.g., breasts), and it is important during the first period of pregnancy. Estrogen is used in hormone therapy. Estrogen, in a pill, patch, or gel form, is the single most effective therapy for suppressing hot flashes.
	o Progesterone	The hormone produced by the corpus luteum during the second half of a woman's cycle. It thickens the lining of the uterus to prepare it to accept the implantation of a fertilized egg. It is released in pulses, so the amount in the bloodstream is not constant. The periodic excretion of estrogen and progesterone during the menstrual cycle is shown below in Figure (76). {{ Tablepic.JPEG-1/8 page}}
	o Inhibin	It is the same one discussed under testes.

1-12 Microbiology

As humans are living longer there is a greater predisposition to infection. This risk is substantially heightened in elderly individuals who are predisposed to infection. The question is whether the microbiological changes that occur within and upon the host influence the process of ageing or is it the biological changes of the host that affects the host's microbiology?

In microbiology, one should know the difference between eukaryotic and prokaryotic cells. Both cells have protective membranes surrounding the cell, This is the only membrane that prokaryotic cells have. The difference is that eukaryotic cells have endomembranes which bind and protect organelles inside the cell such as lysosomes, the endoplasmic reticulum, vesicles, the golgi apparatus, and the mitochondria. Endomembranes are important in the functioning and transportation of substances into and out of the organells and the cells. Prokaryotes are still commonly imagined to be strictly unicellular. Prokaryotic cells do not have a nucleus, and all organells inside the plasma membrane do not have endomembranes, so it is just floating in the cytoplasm. In eukaryotic cells, there are many organells such as mitochondria which convert organic compounds (sugar and fat) to energy (ATP), ribosomes which organize the synthesis of proteins, the rough endoplasmic reticulum which prepares protein for export, the smooth reticulum which regulates calcium level and break down toxic substances and synthesizes steroids, the Golgi which processes and packages substances made by the cell, and the lysosomes digest pathogens. Prokaryotic cells carry out functional processes of life by themselves. They work as a team. They link together to form tissues and organs such as muscle tissue, nerve tissue, and membrane tissue. When these tissues combine together, they can form organs such as the heart, the lung, the liver, etc. Eukaryotes and prokaryotes are organisms that contain all of the enzymes necessary for their life. They stand distinguished from viruses which depend upon the host cell for their life. Life means replication through the process of cell division, and other metabolic functions. Figure (77) shows organells of a prokaryotic cell. A eukaryotic cell has been shown earlier in this chapter.

Figure (77): Prokaryotic cell

98

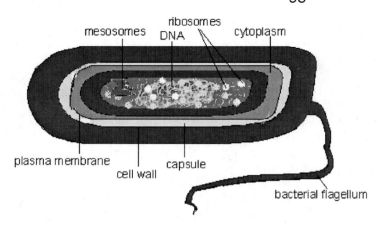

Among the groups of eukaryotic microorganisms are the algae, the protozoa, the fungi, and the slime molds. Bacteria (gram positive and gram negative) are prokaryotes. The cell division in eukaryotes is through the mitosis process where the nucleus divides into two. In prokaryotes, the cell usually divides by binary fission.

1-12-1 Grouping of eukaryotic and prokaryotic cells

Eukaryotes have been broken down into four kingdoms: animals, plants, fungi, and protists or protoctists, Figure (78). The first three kingdoms are a monophyletic group, and the protists kingdom is not monophyletic. A monophyletic group is a group which contains all the descendants of a common ancestor; no monophyletic group contains organisms which are related more to other kingdoms than they are to other protists.

Figure (78): Kingdom of eukaryotes

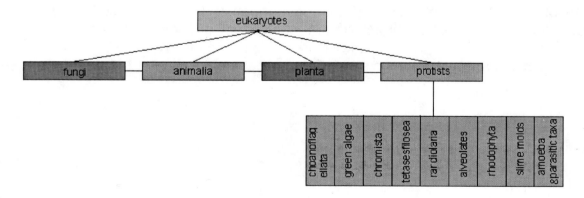

Prokaryotes are mainly divided into two groups (kingdoms); the bacteria and the archaea. The bacteria (the kingdom of monera) in the prokaryotes includes eubacteria (true bacteria), cyanobacteria (blue green algae), and archaebacteria.

All kingdoms of monera do not have a nucleus like other prokaryotes. Figure (79) shows the kingdom of prokaryotes.

Figure (79): Kingdom of prokaryotes

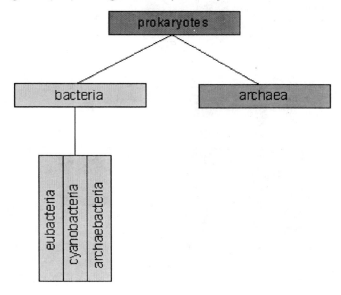

Achaea looks like Figure (80) below.

Figure (80): Archaea

http://en.wikipedia.org/wiki/Archaea

The blue circular shape is not a nucleus as all prokaryotes do not have one.

Figure (81) shows shapes of some eukaryotes

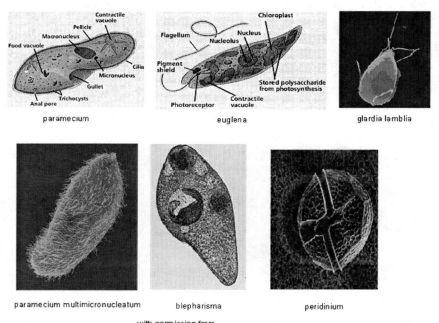

paramecium   euglena   glardia lamblia

paramecium multimicronucleatum   blepharisma   peridinium

with permission from
http://www.emc.maricopa.edu/faculty/farabee/BIOBK/BioBookDiversity_3.html

1-12-2 Growth of bacteria

Bacteria are all around us. They grow in a changing pattern depending on the availability of nutrients, energy sources, temperature, and the pH of the culture medium. A bacterium grows slightly in length and size. The wall of the capsule grows at the centre forming two daughters. At the start of the growth process, bacterium divides at an exponential rate, then at stable rate, and then dies. Figure (82) illustrates the growing pattern of bacteria.

Figure (82): Rate of bacteria growth versus time

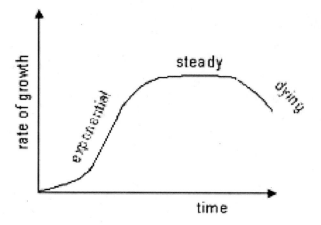

1-12-3 Food microbiology

Many diseases can be transmitted through food. Foodborne illness can be caused by viruses, bacteria, parasites, toxins, metal, and prions. The symptoms range from mild gastroenteritis to life-threatening hepatic, neurologic, renal, and urinal syndromes. Foodborne diseases cause about 80 million cases of illness per year. Some foodborne illness is caused by pathogens that have not yet been identified. Pathogens that cause illness include Listeria monocytogenes, Echerichia coli, Cyclospora cavetanensis, Yersinia etreocotica, Campylobacter, etc.

Table (10) illustrates types of pathogens and agents that cause foodborne illness.

Table (10): Types of pathogens

Pathogens/Agents/ Viruses	Causes and Symptoms
Protozoan parasites	Cause illness in humans and animals through food and water. The disease can be severe and prolonged.
Listeria monocytogens	It is gram-positive bacterial pathogen, caused by contaminated food such as cheese, frankfurters, luncheon meats, raw milk. It is the leading death in adults. Patients must be isolated.
Salmonella	It is transmitted from animals to humans through consumption of undercooked foods and raw eggs. Antibiotics can cause salmonella due to the resistance against their reaction.
Shigella	It is a gram negative species. Symptoms are diarrhea, fever, tenesmus, abdominal pain, seizures, toxic megacolon, arthritis and hemolytic uremic syndrome. It is transmitted by fecal-oral route through food and water. It can spread quickly as it has a powerful mechanism of cell division.
Escherichia coli	It causes E.coli that could cause serious gastrointestinal diseases and death. There are six types of E.coli, most of them cause diarrhea.
Mycotoxins	These are very dangerous species which grow on agriculture food. It could cause chronic diseases, such as cancer in organs such as the liver, heart, kidney, nerves, and stomach. It could damage the immune system and the meiosis process. Monitor food to reduce Mycotoxins.
Bacillus cereus	Six members of B. cereus are food-related disease, causing diarrhea. Deaths have been recorded by types of B.cereus that excrete emetic toxins and CytK toxins. B.cereus is able to grow in refrigerators for a long time, so concern with prolonged shelf life is raised. B. cereus can withstand and survive washing and disinfection (but not UVC and chlorine-based disinfectants).

Yersinia enterocolitica	Pork is the source of Y. enterocolitica. It can grow at a temperature of 0°C. The pathogen can cause reactive arthritis.
Vibrio	Vibrio grows in marine and estuarine media. It is associated with seafood and some types of it could produce a cholera toxin which is an agent of cholera. Vibrio pathogen has the highest rate of fatality (50%) of any foodborne pathogen.
Campylobacter	Campylobacter is one of the major cause of bacterial gastrointestinal disease, but does not cause any complications.
Clostridium botulinum	This pathogen produces extremely potent neurotoxins that result in neuroparalyric disease. Symptoms are diarrhea and intestinal pain. It is caused by food of extended durability in fridges and chilled ready to eat meals.
Staphylococcus aureus	Symptoms include vomiting and diarrhea.

1-13 Food safety

The World Health Organization (WHO), The Department of Food Safety, Zoonoses and Foodborne Diseases (FOS), and the Food and Agriculture Organization (FAO) work together to reduce the serious negative impact of foodborne diseases worldwide. About 1.9 million people die annually from food and waterborne diseases most of whom are children. Many countries have their own codes, acts, standards (ISO), and organizations in order to take care of handling, preparation, and storage of food in better ways in order to prevent foodborne illness.

Bacteria, viruses, and toxins produced by microorganisms are contaminants of food. However, microorganisms can be used to fight these pathogens. Nisin (purified bacteriocins), bacteriophage, and viruses can be added to kill bacterial pathogens. Cooking at certain temperatures can kill bacteria and viruses. However, toxins produced by certain microorganisms may not be affected by heat, and some will not be killed by cooking.

Some organic enzymes and enzymes produced by fermentation can eliminate bacteria and viruses. Yeast which is used for making bread, beer, and wine are less hospitable to bacteria and viruses.

1-14 Biomedical engineering

Biomedical engineering is a topic related to research, design, development, testing, and manufacturing of medical systems and products such as medical instruments, organs, and all medical equipment. Some of the well established specialty areas within the field of biomedical engineering are bioinstrumentation, biomaterials, biomechanics, systems physiology, clinical engineering, and rehabilitation engineering. Biomedical engineering includes:

- Medical devices such as artificial hearts and kidneys, pacemakers, artificial hips and arms, surgical lasers, monitors (sugar, blood pressure, temperature, etc.) and sensors
- Engineered therapies such as prosthesis
- Medical imaging system, and related software
- Computer hardware and software
- Maintainability and reliability of medical instruments
- Training technologists, and technicians
- Bio-nano medical equipment such as nanometers, electronic microscopes, and laser friendly equipment
- Cellular and nucleus engineering equipment used in anatomy, pathology, surgery, repair and regeneration of living tissues and organs for growth of stem cells, and tissue growth in vitro
- Genom and genetic engineering such as mapping and analyzing DNA and other nucleus related organelles
- Equipment for simulating and animating molecular interactions
- Therapeutic engineering used in discovering drugs, and techniques used for sending the drugs to local tissues.

1-15 Psychology and clinical psychology

Psychology is a scientific field involving in the study of mental function and behavior. Psychologists examine behavioral disciplines such as emotion, personality, cognition, and interpersonal relationships.

Sigmund Freud (Austrian from 1890 until 1939) who is still considered the father of psychology developed a method of psychotherapy known as psychoanalysis for the treatment of mental distress psychopathology. He constructed the mind into his well known model 'the tripartite" which consist of the id, ego, and superego. To Freud, id means passion, ego means common sense, and superego means perfection with feeling of guilt.

Freud also tackles the mind through his theories such as the Oedipus complex, free association, and clinical interest in dreams. Anna Freud, his sixth daughter followed his path, focusing on the ego model and its ability to be trained socially. His theories and practice focused on the resolution unconscious conflict often arising from childhood experiences to treat psychological disturbance. However, it was argued that Freud's theories were difficult to test empirically. This led to the so called 'behaviorism' that argued that the contents of the mind were not open to scientific scrutiny and that scientific psychology should only be concerned with the study of observable behavior. There are six major schools of thought on psychology:

- o Structuralism - this was based on introspection, i.e. experimental psychology. This school did not last long. It died with the death of its German founder: Wilhelm Wundt.
- o Psychoanalysis - This school asks for implication of a variety of scientific methods, including experiments, correlation studies, longitudinal studies, and other empirical work to test, measure, explain and predict human behaviour.
- o Functionalism - It was mainly influenced by the evolutionist Charles Darwin. It was based on a more systematic mental process rather than mental consciousness.
- o Behaviorism - It is a theory based on the idea that learning is acquired from the environmental interaction with people who have all different behaviours.
- o Humanism -This theory characterized that all people were created with good instincts, but they deviated from this tendency due to mental and social problems.
- o Cognitivism - It is the branch of psychology that studies that mental process including the way people think, perceive, learn and remember. It also deals with philosophy, neuroscience, and linguistics.

Clinical psychology focuses on the psychological, human adaptation, adjustment, and emotional, biological, social and behavioral function at all environmental conditions. These conditions include, weather, as well as socioeconomic and demographical standards. Clinical psychologists are involved in professional practices that range from the treatment of early minor problems of adjustment to major problems of maladjustment of patients whose disorder requires them to be institutionalized. Clinical psychologists deal with all classes of people (families and organizations) to promote mental health and to alleviate discomfort and maladjustment. They study the two critical relationships: one between brain function and behavior (genetic), and one between the environment and behavior (acquisitioned). Psychological/physiological disorders include the following types, listed in Table (11).

Table (11): Psychological/physiological disorders

A. ADD/ADHD	B. Viable mental disorder and can be treated.
C. Multiple personality disorder	It is called MPD or dissociative identity (DID). It is a condition in which two or more personality states alternately switch a patient's consciousness and behavior. It is not a "split personality"

	disorder.
D. Depression	E. Major depression is known as unipolar depression, and it is a mental disorder characterized by pervasiveness, low self-esteemed, low mood, and loss of pleasure or interest in normally enjoyable activities.
F. Attention deficit hyperactivity disorder	G. Attention Deficit Hyperactivity Disorder (ADHD) is a problem that becomes noticeable in some children in the preschool and early school years. It is hard for these children to control their behaviour and/or pay attention. It can be treated through guidance and understanding from parents, guidance counselors, and the public education system.
H. Anxiety disorder	I. Intense prolonged feelings of fright and distress can cause anxiety disorder. Patients with anxiety disorders could turn their life into a fearful and uneasy state, and could negatively affect their relationships with family, colleagues and friends. Some signs of anxiety disorder are: o Panic disorder, which is panic attacks that strike without warning, and is accompanied by sudden feelings of terror. Physically, an attack may cause chest pain, heart palpitations, dizziness, shortness of breath, abdominal discomfort, and feelings of worthlessness and fear of dying.

	<ul><li>Phobias - Phobias are categorized into two categories: specific phobias, such as fear of flying, blood, heights and open spaces, and social phobia, which involves fear of social situations such as a paralyzing, unreasonable self-consciousness about social meeting or gathering.</li><li>Obsessive-Compulsive Disorder (OCD) - is a disorder in which people suffer from persistent unwanted thoughts and practices (obsessions such as touching the lips very often, or turning the light on and off many times before leaving the room) and / or rituals (compulsions) which are impossible to control. Typically, obsessions concern contamination, doubting (such as worrying that the door has not been locked) and disturbing sexual or religious thoughts. Compulsions include reassuring washing, checking, and counting.</li><li>Post-Traumatic Stress Disorder - A terrifying incident in which serious physical harm occurred or was threatened can cause post-traumatic stress disorder. Survivors of war, rape, child abuse, or a natural disaster may develop post-traumatic stress disorder. Common symptoms include flashbacks, during which the person re-lives the terrifying experience, nightmares, irritability, and a bad temper.</li></ul>
J. Heterosexism and homophobia	Like other forms of discrimination and harassment, homophobia and heterosexism weakens the diversity of genders. In order to increase

	awareness of homophobia and heterosexism, it is useful to be aware of sexuality: gender identity and sexual orientation. K.
L. Schizophrenia	Symptoms are: o delusions and/or hallucinations, o lack of motivation, o social withdrawal, o Thought disorders. Paranoia and phobias are cause of schizophrenia
M. Bipolar disorder	N. Symptoms are: o extreme irritability, o feelings of euphoria, extreme optimism, exaggerated self-esteem, o rapid speech, racing thoughts, o impulsive and potentially reckless behavior. O. It can be treated.
P. Anorexia nervosa	Anorexia nervosa is an eating disorder affecting mainly girls or women, although boys or men can also suffer from it. The cause of anorexia nervosa is unknown. People in certain professions, such as modeling and ballet dancing, are especially at risk. It can be treated.
Q. Alcoholism	R. Social disruption, failure to meet obligations, disputes at work and personal relationships, poor work production and quality, hypertension, low performance in school, and

	family problems are causes of alcoholism.
S. Sleep disorder	T. Sleep disorder can be divided into 6 types: ○ Sleep apnea is a common disorder in which one has one or more pauses in breathing or shallow breaths while you sleep. ○ Insomnia is a difficulty falling asleep, staying asleep, or not feeling refreshed after 6 to 9 hours of being in bed. ○ Narcolepsy is characterized by excessive sleep which leads to disturbed nocturnal sleep. This can be confused with insomnia. ○ Restless legs syndrome is a neurological disorder characterized by unpleasant sensations in the leg (and sometimes in the arms), as if something is burning, creeping, tugging or crawling inside the leg. ○ Shift work sleep disorders affect employees who work regular evening or night shifts that require changing sleep patterns. This pattern of disorder, for unknown reasons, is associated with some types of cancers such as breast cancer.

1-16 Physiology

As we age, we undergo a number of physiological changes, which affect not only how we look, but also how we function and respond to daily living. Overall, the changes in the later life span involve a general slowing down of all organ systems due to a gradual decline in cellular activity.

Physiology (the study of the functioning of organisms) and anatomy (the study of structure of the body) are tied together because each represents a feedback signal which corrects the other if there is a problem with the function of the cells, organs, or the biochemistry of a body. The signal can detect the temperature, the water, the pressure, and/or the pH level in order to correct the deviance of the body system. These deviances may occur witin the nervous system, the respiratory system, the integumentary system, etc. Figure (83) represents the interaction between the physiology and anatomy of the human body. In figure (83), pH stands for acidity and alkalinity, T for temperature, L for liquidity, and P for pressure. If any one of the variables (pH, T, L, or P) or other environmental factors, such as the level of contamination, has been changed, the system will change its normal function. This would lead to a system disturbance called pathophysiology.The body will react to reverse the change and correct its homeostasis (stability and constant condition).

Figure (83): Homeostasis is maintained through feedback modification to the changes in variables.

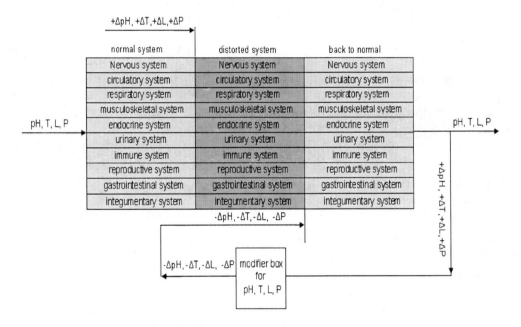

1-17 Pharmacology and toxicology

Toxicology-induced damage can deteriorate cell cycle, DNA repair mechanism, functions of mitochondria, and can cause DNA damage in apoptosis, deafness, neurodegenerative disease and other diseases of aging.

Pharmacology is different from pharmacy because pharmacology is the study of the drugs and their actions from the molecular level through the whole body. It deals with the interaction between the cell receptors and the drug molecule, where as pharmacy deals with a drug's structure and reactions between chemicals. Pharmacology is a multi-disciplinary science with many subspecialties , including cardiovascular, clinical, neuropsycal, behavioral pharmacology, pharmacogenetics, pharmaeconomics, pharmakinetics, etc. Toxicology is similar to pharmacology because both require an understanding of basic knowledge of the properties and reactions of chemicals. Toxicology puts emphasis on the adverse effect of chemicals and risks of assessment while pharmacology focuses on the therapeutic effect of drugs and chemicals.

Pharmacology deals with five main subjects:

1-17-1 Neuropsyco pharmacology

Parkinson's disease (PD) is a degenerative disorder of the cell's nature and science that damages the motor skills, speech and other brain functions. It is characterized by muscle rigidity, loss of physical movement, tremors, loss of speech, and, in extreme cases, loss of memory. Researchers reported that the cause of Parkinson's disease is the insufficient formation of dopamine, which is produced by dopamenergic neurons of the brain. Patients with Parkinson's disease may suffer cognitive dysfunction and subtle language problems.

Epilepsy is a brain disorder in which clusters of nerve cells, or neurons in the brain sometimes signal abnormally. The neural system of the brain becomes disturbed and malfunctions, causing strange sensations, emotions, and behaviours. Sometimes this causes tremors, muscle spasms, and loss of consciousness. Epilepsy may develop because of a malfunction in the brain wiring and switching, an imbalance of nerve signaling chemicals called neurotransmitters, or some combination of these factors. Seizures are accompanied with epilepsy.

Psychosis is a serious but treatable disorder. People with psychosis may have hallucinations, loss contacts with reality, bizarre behaviour, and delusional thinking. A person experiencing psychosis can overwhelm the lives of families and individuals.

Antimicrobial chemotherapy is a therapy that targets the microorganism but not the cell. When chemotherapy is used, there is a margin between the chemotherapy's actions on the pathogen and the effect of toxity (produced due to the application of chemotherapy) on the host cells. Such a margin must be broad that the toxin does not interfere with the effect of chemotherapy. Table (12) shows drug names and uses against bacteria, viruses, fungi, and protozoa organisms, with permission from: http://www.microbiologybytes.com/iandi/9a.html

Table (12): Drug names against bacteria

Antibacterial drugs (common)	Uses
Penicillin	Pneumococcal. Meningococcal, streptococcal infections
Tetracycline	Chlamydial, mycoplasmal infections
Monobactams	Pseudomonas
Macrolides	Gram-positive cocci, legionells, Chlamydia
Cephalosporin	Respiratory and urinary infections
Quinolones	pseudomonas
Vanomycin	MRSA
Sulphonamides	In co-trimoxazole
Aminoglycosides	Gram-negative coliform infections
Chloramphenicol	Salmonella and haemophilus influenza infections
Nitroimidazoles	Anaerobic infections
Antiviral drugs (common)	Uses
Azidothymidine	HIV
Nevirapine	HIV
Ribaverin	Respiratory syncytial virus (RS-virus)
Gansiclovir	Cytomegalovirus
Rimantadine, amantadine	Influenza A
Acyclovir, famciclovir, valaciciovir	Herpes virus
Antiprotozoal drugs (common)	Uses
Chloroquine	Malaria
Pyrimathamine	Malaria, Toxoplasma
Pentamedine	Gambiense, rhodesiense, trypanosoma, pneumocystis
Imidazoles	Entamoeba, glardia, Trichomonas
Antifungal drugs (common)	Uses
Flucytosine	Serious fungal infection
Amphotoricin B	Systemic fungal infections
Azoles, clotrimazole, miconazole, fluconazole	Local candidosis and dermatophyte infections

1-17-2 Antiviral chemotherapy

Some viruses are not completely amenable to vaccines and drugs. The only two drugs proved their success against viruses (influenza, herpes viruses, rhinoviruses, arboviruses) if they are combined with other antiviral drugs, such as acyclovir and AZT. Even with acyclovir and AZT, the combination does not have multispectral activity. Antiviral chemotherapy, in general, is made of:

a. Pyrophosphate analogues, which have two atoms of phosphorus and two double bonded oxygen atoms. The two atoms of phosphorous inhibit the DNA of some types of viruses, such as herpes viruses and HIV due to the negative oxygen (anion) bonded to phosphorous atom, Figure (84).

Figure (84): Pyrophosphate (2P) and virus DNA become triphosphate and inhibit the virus DNA.

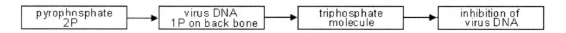

b. Nucleoside analogues to which one or two atoms can be added, will change the shape of the virus DNA. Nucleosides can be added to phosphorous atoms by specific kinases (1P and 4 oxygen's, two of them are negatives) in the cell on the sugar, replacing the third hydrogen atom. This produces a nucleotide which is the building block of the DNA and RNA. Thus, it blocks viruses such as the herb virus and HIV.
c. Amantidine molecules are hydrocarbon molecules in which one group of the four methanes is replaced by an amino group (NH_2). They are used against Influenza A (Asian influenza) infections. Amantidine is a drug used to prevent the entry of viruses to the brain. It is used to help reduce symptoms of Parkinson's disease.

1-17-3 Cancer chemotherapy

Cancer chemotherapy is a method of treating cancer by using chemical drugs or a combination of drugs. Chemotherapy is meant to kill, slow or stop cancer cells from growing, spreading, or multiplying to other organs of the body.

Cancers are made of billions of cells few of which can travel from one part of the body to the other, causing cancer in that part. The new cancer is known as "metastasis". The few travelled cells cannot be detected even with the most sophisticated scanners and blood tests. Such metastatic cancers can grow, and ultimately cause incurable cancer. The chemotherapy is used to reduce the number of these travelled cells, and to reduce the number of reocurrences.

Chemotherapy can damage healthy cells in addition to cancerous cells, thus, it causes many side effects. The body's healthy cells usually fix (repair) the damaged healthy cells after a period of rest from the chemotherapy. Some other drugs are given to the patient in order to reduce or enhance the side effects. Cancer cells are more sensitive to chemotherapy than healthy cells because they divide faster and they divide in an uncontrollable way. Healthy cells, (especially the rapidly dividing cells of the skin, the lining of the stomach and intestines, and the bladder), can also be affected in a rapid fashion. Some of the side effects of chemotherapy are:

- Blood side effect - the blood cells (RBC, WBC, and platelets) are the most rapidly dividing cells, therefore, they are the most sensitive to chemotherapy. The body's blood count can decrease with chemotherapy to the lowest level termed as" nadir". When RBCs decrease significantly, a condition called "anemia" occurs, and patients feel tired and short of breath. A blood transfusion may be required. If a significant decrease in WBC called 'neutropenia", occurs it would be difficult to fight infections. Neutropenia is accompanied by fever, chills, coughing, sore throat, and skin and mouth rashes. The decrease in platelets causes "thrompocytopenin", which causes easy bruising and recessive bleeding.
- Hair loss is a side effect of chemotherapy known as "Alopecia areata".
- A sore throat is called "Stomatitis" which makes food swallowing difficult and painful.
- Fertility and sexuality may affect sperm count and viability. Women should take birth control before taking measures since chemotherapy is very toxic in some cases, and it also could change the psychological mood of women of childbearing age due to the change of the amount and timing of excreted estrogen and progesterone. Chemotherapy may change the menstrual cycle, and could result in total absence of periods. Dryness of the vagina and increased pain with sexual intercourse are also the result of chemotherapy.
- Other side effect include diarrhea, constipation, rashes on the skin, and some types of allergies which can be the effect of chemotherapy.

Chemotherapy does not guarantee a cure because too many patients relapses after treatment. The use of chemotherapy depends on the risks and benefits of the patient as determined by the doctor.

1-18 Human brain

Researchers at Children's Hospital Boston and Harvard Medical School have uncovered a kind of genetic signature associated with the aging human brain that may contribute to cognitive decline associated with aging. One of the study's more surprising results was that these gene changes start in the 40s for some individuals. The results raise intriguing questions about when and why the brain begins to age and the possibility of developing strategies to protect critical genes early in life in an attempt to preserve brain function and delay the onset of age-related conditions such as Alzheimer's disease.

The human brain is a complex organ that controls our body, receives and transmits information, and analyzes and stores information. It is the organ that allows human to see, hear, smell, think, taste, move, and feel. The brain generates electrical signals that interact with chemical particles in the body. The result is transmitted through wires (nerves) to motor other organs as required. Figure (85) represents a simulation between the brain (generator) and the body's organs.

Figure (85): Feedback signal of correlation between brain and organs.

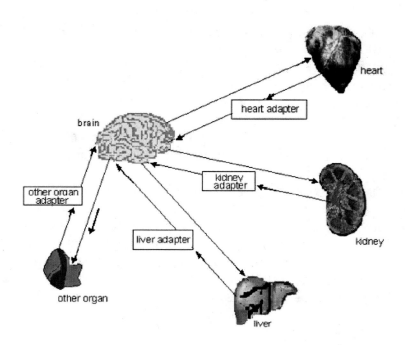

The brain consists of white matter (60%) and gray matter (40%) which is contained within the skull. Brain cells include neurons and glial cells (Glial cells provide support and protection for neurons, they are somewhat like glue material).

The brain has the following parts: the frontal lobe, the central sulcus, parietal lobe, the occipital lobe, The cerebellum, the medulla oblongata, the temporal lobe, and the lateral sulcus, Figure (86).

Figure (86): Exterior left side of the cerebrum

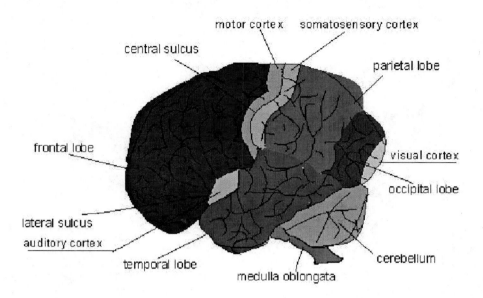

The brain weight is about 2% of the body's weight, but it uses 20% of the oxygen supply. If brain cells do not get oxygen for 3 to 5 minutes, they begin to die.

The three main parts of the brain are:

- o The cerebrum: it fills up most of the skill. It is responsible for solving problems, remembering, feeling thinking, and it controls movement.
- o The cerebellum: It is located at the back of the head, under the cerebrum. It controls balance and coordination.
- o The brain stem: is beneath the cerebrum, and in front of the cerebellum. It connects the brain to the spinal cord, It controls breathing, digestion, blood pressure, and heart beat rate.

1-18-1 Frontal lobe

The frontal lobe is known as the motor cortex and the parietal lobe is known as the sensory cortex. It contains the dopamine-sensitive neurons which are responsible for attention, long-term memory, planning and drive. Dopamine can affect the function of the modal activity causing disorders such as schizophrenia. As a chemical messenger, dopamine is similar to adrenaline because it controls emotional response, pleasure and pain, and physical movement. In Parkinson's disease, the brain does not produce dopamine. Patients with Parkinson's disease are given L-Dopa, a drug that is converted to dopamine inside the brain. The frontal lobe also contains myelin which is a connecting cluster of wires (fibers of electrically-insulating dielectric components) that control and transmit signals from Soma cells to the axon terminal, Figure (87).

Figure (87): Spikes movement from spine head or spike neck to interneurons or vice versa

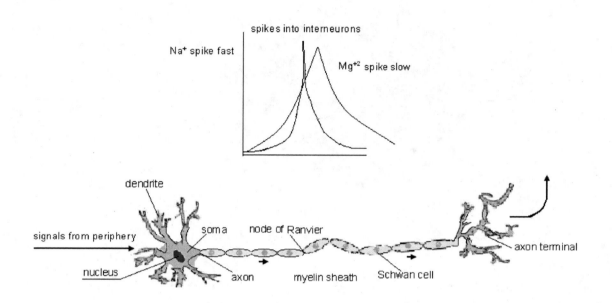

1-18-2 Central sulcus

The central sulcus is a deep groove in the brain that separates the frontal lobe and parietal lobe of the cerebrum. It has a 'map' of the human body on each side that corresponds to the other side. Mapping of the human body is called " homunculus". So when the sensory part of the parietal lobe is stimulated by any change in the electron charge, say more sodium in the body; Na^+, its associated motor part on both sides of the body will respond. It will do this by decreasing the sodium through a signal go to the kidney. If the right hand feels itchiness, the parietal lobe will react and sends a signal to the motor cortex of the frontal lobe for correction and elimination of itchiness by excreting the appropriate peptide (hormone).

1-18-3 Motor cortex

In 1870, Hitzig and Fritch electrically stimulated various parts of the dog's motor cortex. They found that each stimulated part (spot) of the motor cortex was responsible for a certain part of the body. For example, the left hand represents the xth spot on the motor cortex, the right hand corresponds to y^{th} spot, and so on. They destroyed some spots of the cortex, and found that the corresponding part of the body became paralyzed. This is how it was discovered that every part of the body has a particular region of the primary motor cortex that controls its movement. But what are remarkable about this "motor map" are that certain parts

of the body—those that can make the finest movements—take up much more space than others,
http://thebrain.mcgill.ca/flash/d/d_06/d_06_cr/d_06_cr_mou/d_06_cr_mou.html

Figure (88): Length of parts of the motor cortex is inversely proportional to the size of the organ.

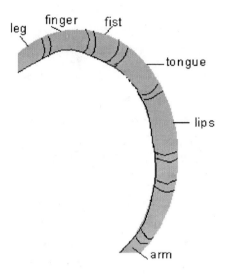

1-18-4 Somatosensory cortex

Neurons in the somatosensory cortex are activated when skin is touched. Somatosensory cortex is divided into portions; each of them is responsible for a specific organ of the body, Figure (89).

Figure (89): Parts of somatosensory cortex; in charge of touches

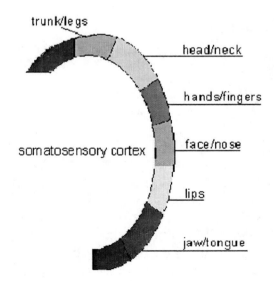

1-18-5 Parietal lobe

The parietal lobes can be divided into two functional areas. One is responsible for sensation and perception, and the other is responsible for integrating and coordinating sensory input which forms a single perception; cognition. Damage to the parietal lobe could lead to abnormalities in body image and spatial relations (Kindel, Scwartz, and Jessel, 1991).

Damage to the left side of the parietal lobe could result in right-left confusion such as difficulty in writing (agraphia), difficulty in mathematics (acalculia), difficulty in language assembly and composition (aphasia), and inability to recognize and identify objects (agnosia).

Damage to the right side of the parietal lobe could result in ignoring daily and routinely jobs such as shaving, washing, and dressing. It could results in disorganization and deficiency in doing physical work such as painting, drawing, singing, etc.

1-18-6 Visual cortex

Visual cortex has mainly five visual areas; V1, V2, V3, V4, and MT. Nearly all visual information reaches the cortex via V1, the largest and most important visual cortical areas, Figure (90).

Figure (90): Main five areas of the visual cortex

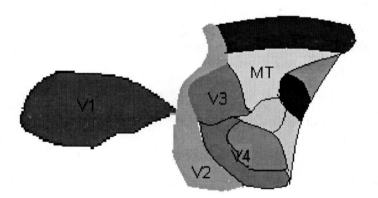

The axons of the ganglion cells in the retina collect in a bundle of verves at the optic disc and come out from the retina to form the optic nerve. The optic nerve carries the impulses seen by the eyes to the lateral geniculate nucleus (LGN), and then to V1 in the visual cortex. The vast majority of the nerve fibers tract projected light to the lateral geniculate nucleus (LGN) in the dorsal part of the thalamus. Any damage at any point along the pathway from the retina to the visual cortex results in blindness or semi blindness. The primary pulses go to the primary visual cortex V1 which analyzes most of the outside picture, but the visual cortex sees the real picture in a fussy shape. The primary cortex sends the pulses to the secondary visual cortex (V2) for a better picture. The process is interlinked between V1 and other sub branches of the visual cortex until the picture becomes as it is seen. The sub branches of the cortex have distinctive traits of responding to far more complex shapes, Figure (91).

Figure (91): Interlink between eyes and visual lobe

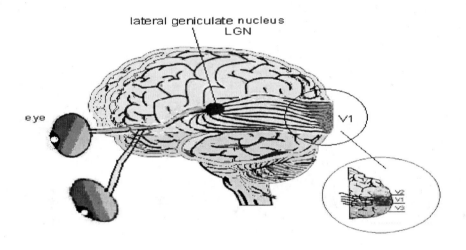

1-18-7 Occipital lobes

The occipital lobes are the smallest lobe in the human cerebral cortex. They are the centre of our visual perception of the field of vision, and located at the back of the brain, thus they are invulnerable to injury. However, any significant damage to the brain could lead to problems in the visual system such as visual field defects and scotomas which is a circular black point surrounding the center of the field of vision. The occipital lobe is involved in visuospatial processing, discrimination of movement and color discrimination (Westmoreland et al., 1994).

Retinal sensors have two sides; the outside and the inside, and both are connected to optic nerves which go to occipital lobes through the LGN. The occipital lobes (two) are located on both sides of the brain. Each occipital lobe receives sensory pulses from the outside half of the retina on the same side of the head, and from the inside half of the retina on the other side of the head, Figure (92).

Figure (92): Connection between retina and occipital lobes

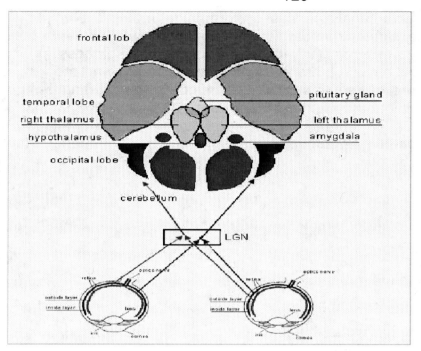

1-18-8 Cerebellum

Cerebellum is part of the brain that is involved in the coordination of voluntary motor movement, balance and stability (equilibrium) and muscle tone. Cerebral damage could lead to many defects in physical responses such as: ataxic (sway and staggering in walking), asynergia (loss of coordination of motor movement), nystagmus (abnormal eye movement), dysmetria (inability to know distances), hypotonia (weak muscle), ataxic dysarthria (slurred speech), etc.

1-18-9 Hypothalamus

One of the main functions of the hypothalamus is to link the nervous system to the endocrine system through the pituitary gland. In humans, it is roughly the size of an almond. It controls all endocrine glands in the body, and regulate blood pressure (through vasopressin and vasoconstriction), body temperature, pH level, metabolism (through TSH), and adrenaline levels (through ACTH). The main functions of the hypothalamus are:

- o retina - some fibers from the retina optic nerve go straight to a small nucleus within the hypothalamus called the suprachiasmatic nucleus. This nucleus controls circadian clock and rhythm, and coordinate it with the light/dark cycles.
- o Endocrine signals (hormones of chemical molecules) from posterior pituitary gland's receptors are sent via axons to the hypothalamus. For example, oxytocin hormones (a pituitary hormone that stimulates uterine contractions during childbirth and triggers lactation) are released from the

exterior pituitary gland into the axons to be processed into the hypothalamus and then to the blood stream to induce labor.

- o Limbic and olfactory systems - The system contains parts such as the amygdala (believed to have the primary role in the processing of memory), the hippocampus (plays a major role in short term memory and spatial navigation), the olfactory cortex (responsible for the sense of smell), and anterior thalamic nuclei project to the hypothalamus. In addition to above functions, they help to regulate behaviors such as eating and reproduction.
- o Nucleus of the solitary tract - this nucleus conveys afferent (afferent means upwards, efferent means downwards) information from receptors in the walls of the cardiovascular, respiratory, and intestinal tracts to the hypothalamus and other targets. Information includes blood pressure control and gut expansion and contraction.
- o Circumventricular organs - these nuclei are located along the ventricles, and are unique in the brain in that they lack a blood-brain barrier. These organs secrete or are sites of action of a variety of different hormones, cytokines and neurotransmitters. They monitor substances in the blood that would normally be shielded from neural tissue, such as OVLT (Organum Vasculosum Laminae Terminalis, and are responsible for osmotic blood pressure) which is sensitive to changes in osmolarity, and the area postrema, which is sensitive to toxins in the blood and can induce vomiting. Both of these relay to the hypothalamus.
- o Neural signals to the autonomic system - These activities are generally performed without conscious control (involuntary). Neuron signals affects heart rate/ vasoconstriction, digestion, respiration rate, salivation, perspiration, diameter of the pupils, urination, sweating and sexual responses. The (lateral) hypothalamus projects to the (lateral) medulla, where the cells that drive the autonomic systems are located.
- o Reticular formation - The reticular plays an important role in sleep and consciousness as well as modulation of pain, alertness, fatigue, and motivation to perform various activities. It also controls activities such as specific behaviors, including eating, urination, defecation, walking, sexual activities, skin temperature, which all project and relayed to the hypothalamus, http://en.wikipedia.org/wiki/Reticular_formation, http://thalamus.wustl.edu/course/hypoANS.html.

CHAPTER   2

CHEMISTRY

Aging is characterized by gradual changes in the structure, function and turnover of proteins, sugar, lipids, enzymes, hormones, minerals, etc. The rate of these changes is regulated, in part, by hormonal mechanisms. This chapter will address action and reaction of atoms and molecules inside human body including oxidatation and reduction that cause diseases. This chapter includes elucidating the the reaction kinetics, ATP, and reaction between atoms. The free radicals theory probably arises largely through reactions involving molecular oxygen catalyzed in the cell by oxidative enzymes and in the connective tissues by traces of metals such as iron, cobalt, and manganese. Chemistry can emerge into:

- Definition of oxidative damage of reactive oxygen and nitrogen
- Shaping of interaction of proteins, lipids and DNA (both genomic and mitochondrial) with aging in different tissues and organs
- Devoloping of relationship between cell signaling pathways and aging, for example, how to boost homones (say dopamine in brain) to avoid the risk of Alzheimer disease.
- Determining endogenously-generated xenobiotics (chemical substances that are forein to the biological system)
- Generating chemical molecules (drugs) that can alter the declining of life span due to aging.

Chemistry is the science that deals with the compositions and properties of matters. Understanding of chemistry and chemical reaction helps us to comprehend the normal and abnormal functioning of our body. Before we involve into the detailed chemical activity of the body, we provide some insights into the fundamental forms of matter such as atoms, molecules and elements:

2-1 Atoms

The atom is the smallest unit of an element that holds the chemical and physical properties of that element. An atom has an electron cloud consisting of negatively charged electrons orbiting a dense nucleus. The nucleus contains positively charged proton and electrically neutral neutrons, Figure (1-1).

Figure (1): The atom

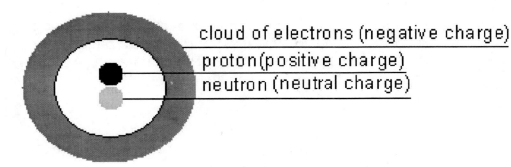

cloud of electrons (negative charge)
proton(positive charge)
neutron (neutral charge)

Matter is composed of molecules, and molecules are composed of atoms. This is a consequence of the manner in which the electrons are distributed throughout space in the attractive field exerted by the nuclei. The nuclei act as the center of attraction of a cloud of negative charge. The electron density describes the manner in which the electronic charge is shaped in the real space. The electron density and the direction of rotation shape the characteristic property and determine the appearance and form of matter.

When the number of protons in the nucleus equals the number of electrons, the atom is electrically neutral and stable. When there is a difference in numbers between protons and electrons, then it is called "ionic atom' which carries a positive or negative charge. An atom is classified according to its number of protons and neutrons: the number of protons determines the type and the name of chemical element and the number of neutrons determines the isotope of that element.

2-2 Molecules

Molecule can found as a unit of two or more atoms held together by covalent bonds or ionic bonds. We shall discuss both types of bonds, but will begin with the binding of atoms. Atom is like a ball with three dimensions. So the attraction and repulsion between atoms are accomplished also in multidimensional manner. Let us consider the molecule of the ethylene Gas C2H4. It has the molecular structure of H2C=CH2 and it can be shown in a Planner form of figure (2):

Figure (2): Planner chemical form of ethylene molecule

Figure (3-a) shows the electron density in the plane containing the two carbon and four hydrogen nuclei of the molecule ethylene, portrayed as a projection in the third dimension and in the form of a contour map. The actual density of carbon atom is much larger than the one shown in the picture, http://www.hec.utah.edu/anions/index_files/image

Figure (3-a): Electron density of ethylene

Figure (3-b) is the same as in Figure (3-a) but for a plane obtained by a rotation of 90° about the C-C axis, a plane containing only the carbon nuclei.

Figure (3-b): Electron density of ethylene rotated 90°

Figure (3c) is the same portrayal as in Figure (3-a), but this time for a plane perpendicular to the C-C axis at its mid-point.

Figure (3c): Electron density perpendicular to figure (3-a)

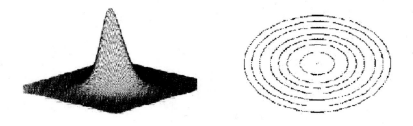

The three pictures of the molecule corresponding to above three figures; Figure (3-a), (3-b), and (3-c) are shown in Figure (4-1) below.

Figure (4): Envelope of the electron density for the molecule ethylene

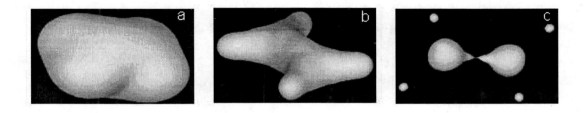

2-3 Bonding of atoms

There are mainly two types of the bonding of atoms: covalent and ionic bonding.

A covalent bond is a form of chemical bonding that is distinguished by the sharing of pairs of electrons between atoms. The stability of covalent bonds is based on the attraction-to-repulsion forces between atoms. Let's take the following examples of methane, nitrate, and water as shown in Figure (5).

Figure (5): Covalent bond of methane, nitrate, and water

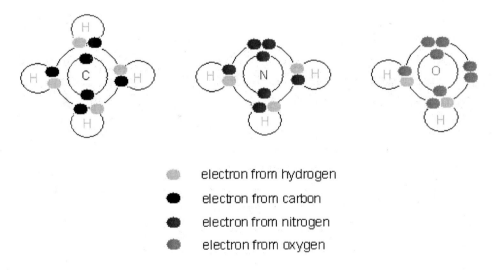

    electron from hydrogen
    electron from carbon
    electron from nitrogen
    electron from oxygen

An ionic bond is an electrical attraction between two oppositely charged atoms or groups of atoms. All atoms in molecules are trying to have their outer orbit in a stable condition or neutrality, i.e. 8 electrons. In order to gain stability, they will either lose one or more of its outermost electrons thus becoming a positive ion (cation) or they will gain one or more electrons thus becoming a negative ion (anion). Therefore, atoms of negative charges attract with those of positive charges. That electrical attraction between two oppositely charged ions is referred to as an ionic bond. Most salts are ionic. Any metal will combine chemically with any non-metal to form ionic bonds that hold the molecule together.

Let's see the ionic bond in the reaction of sodium with chlorine.

Chlorine (on the right) pulls one valence electron from sodium atom, Figure (6) below.

Figure (6): Ionic bond of sodium chloride

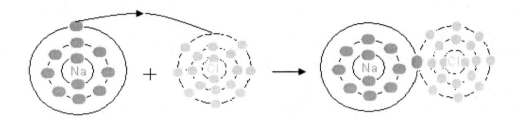

The outcome is the table salt (sodium chloride) as below:

Note that the valance in the chlorine is 7 (high negative), and the valance in the sodium is 1 (low negative), therefore, the chlorine pulls the one electron from sodium.

2-4 Valence-Shell Electron-Pair Repulsion Theory (VSEPR)

The electron density distribution for a molecule is best illustrated by means of a contour map. The contour map of the charge distribution for the lowest or most stable state of the hydrogen molecule is show in Figure (7) which shows a cross section of one molecule of the hydrogen,
http://www.chemistry.mcmaster.ca/esam/Chapter.

Figure (7): Charge distribution of hydrogen molecule

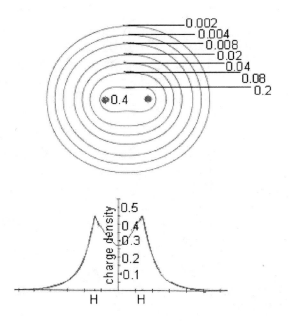

Another example of helium molecule is shown in Figure (8).

Figure (8): Repulsion between the two atoms of the helium molecular

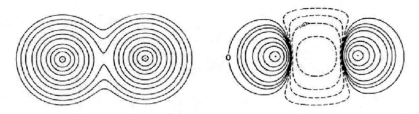

The VSEPR theory assumes that each atom in a molecule will achieve a geometry that minimizes the repulsion between electrons in the valence shell of that atom.

The Valence Shell Electron Pair Repulsion (VSEPR) theory allows us to predict the 3-dimensional shape of molecules from knowledge of their Lewis Dot structure (it will be explained). In VSEPR theory, the position of bound pairs and lone pairs are described relative to the central of the atom. Once the bound and lone pair electrons are positioned, the resulting geometrical (physical) shape presented by the atoms is only used to describe the molecule.

2-5 Lewis structures of atoms

The chemical symbol for the atom is surrounded by a number of dots corresponding to the number of valence electrons. Lewis Structures for ions of elements are shown in Table (1).

Table (1): Lewis structure for atoms

charge on ion	1+		2+	3+	4+	4-	3-	2-	1-	
electrons lost or gained	1e lost		2e lost	3e lost	4e lost	4e gained	3e gained	2e gained	1e gained	
group lost or gained	H+	group 1 1+ alkali metals	group II 2+ alkali earth metals	group III 3+	group IV 4+	group IV- 4-	group V 3-	group VI 2-	group VII 1- halogen	H- hydride
Lewis structure electron dot diagram	H^+	Li^+	Be^{2+}	B^{3+}	C^{4+}	C^4	N^{3-}	O^2	Fl^{1-}	

2-6 Lewis structure for Ions, Table (2)

Table (2): Lewis structure for ions

Number of valence electrons	1		2		3	4	5	6	7	8
group	hydrogen	group I alkali metals	helium	group II alkali earth metals	group III	group IV	group V	group VI	group VII	group VIII
Lewis structure electron diagram	H	Li	He	Be	B	C	N	O	F	Ne

Let us take the molecule of beryllium chloride, $BeCl_2$. The electronegativity difference between beryllium and chlorine isn't enough to allow the formation of ions as there are 4 electrons forming 2 pairs balancing each other.

Beryllium has 2 outer electrons because it is in group 2. It forms bonds to two chlorines, each of which adds another electron to the outer level of the beryllium..

The two bonding pairs arrange themselves at 180° to each other, because it's as far apart as they can get. The molecule is described as being linear as shown in Figure (9).

Figure (9): Linear bond

Before we proceed in different type of molecule bonding, we should always remember that:

Greatest repulsion		lone pair - lone pair
Medium repulsion		lone pair - bond pair
least repulsion		bond pair - bond pair

Bonds of different shapes are shown in Figure (10).

Figure (10): Bonds of different shapes

This Molecule of boron trifluoride (BF_3) has 3 electrons in its outside orbit. They are pulled by the chlorine atom which has 7 electrons in its outside orbit. Note that boron in group 3 (13) and chlorine in group 7 (17) of periodic table. The outcome of attraction and repulsion is 3 bonding pairs which are at 120° to each other as shown in the corresponding plan.

This methane molecule (CH_4) has four bonding pairs arranged in a tetrahedral global arrangement. The angle between two hydrogen atoms is 109.5° calculated based on the central gravity of all of atoms that looks like regular three triangles with a based pyramid.

This ammonia molecule (NH3) has the nitrogen in the centre, and the angle between two hydrogen atoms is 107.8° as shown. The molecule is arranged in two lone pairs and one single electron. One of the two pairs is bonded to the two hydrogen atoms. The ammonia molecule has one lone pair and one bonding pair. The bonding pair is longer than the lone pair. Therefore, there is more repulsion between them than the two bonding pairs.

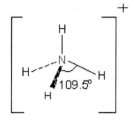

This ammonium ion (NH_4^+) has one hydrogen atom more than the normal capacity of three electrons in the nitrogen atom far most orbit. If the number of hydrogen atoms attached to the nitrogen atom is 3, the ammonium atom is neutral. The molecule is now positive as it carries one more hydrogen atom (proton). The ammonium molecule gains one electron from the orbit of the hydrogen atom, and in the same time, it gets one positive proton. Because the proton positivity is much larger than the negativity of the electron, therefore the ammonium molecule is positive.

Physical shapes of some molecules are shown in figure (11).

Figure (11): physical shapes of some molecules

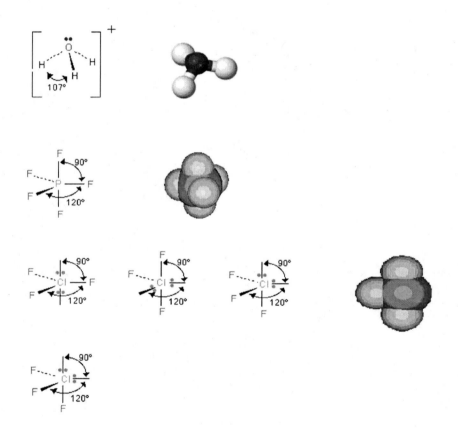

Lewis dot diagram is a powerful tool of covalent and ionic molecules as shown in Figure (12).

Figure (12): Lewis dot diagram of covalent and ionic molecules

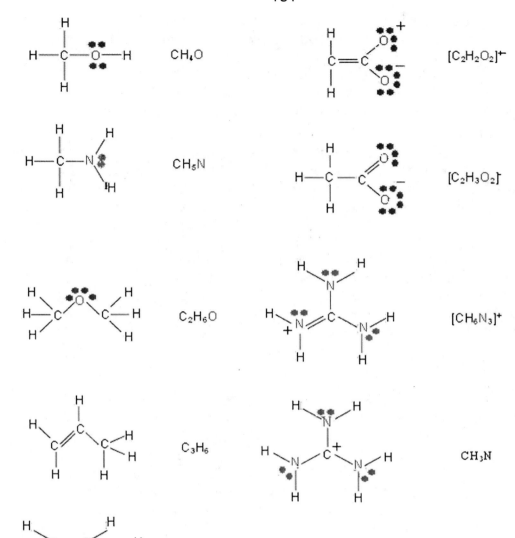

2-7 Components of the cellular life

2-7-1 Water

- Comprises 60 - 90% of most living organisms (and cells).
- Important because it serves as an excellent solvent and involves into most metabolic reactions.

2-7-2 Carbohydrates

The carbohydrates help in regulating the metabolism and give more strength to the body. The carbohydrates give good growth as well. Since the food that contains carbohydrates gets digested soon people tend to eat more after the digestion. This is the only problem with the carbohydrates food. The carbohydrates are found in the liver and the muscles in the form of glycogen. Consumption of food that contains carbohydrates helps in quick digestion and it is a best secret agent for the Anti aging process, see chapter 3.

As the name implies, a carbohydrate is a molecule whose molecular formula can be expressed in terms of just carbon, oxygen and hydrogen and takes the formula of CxHyOz. If one hydrogen atom is lost the sugar becomes an aldehyde (in case of glucose) and is termed an aldose, or it is a ketone (in case of fructose) and is termed a ketose (combination of glucose and fructose is sugar, and will be discussed later). Monosaccharide contains either a ketone or aldehyde functional group, and hydroxyl groups on most or all of the non-carbonyl carbon atoms. For example, glucose has the formula $C_6H_{12}O_6$ and sucrose (table sugar) has the formula $C_{12}H_{22}O_{11}$. More complex carbohydrates such as starch and cellulose are polymers of glucose. Their formulas can be expressed as $C_n(H_2O)_{n-1}$ or $C_n(water)_{n-1}$.

2-7-2-1 Types of carbohydrates:

- Monosaccharide (e.g., glucose or fructose) - most contain 5 or 6 carbon atoms
- Disaccharide consists of 2 monosaccharide linked together.

Examples include sucrose (a common plant disaccharide is composed of the monosaccharide glucose and fructose), as shown in Figure (13).

Figure (13): Glucose and fructose together (sugar), each of them has $C_6H_{12}O_6$ formula.

In linear form, two monosaccharide (glucose and fructose) linked together to form sucrose, Figure (14).

Figure (14): Glucose and fructose in linear form

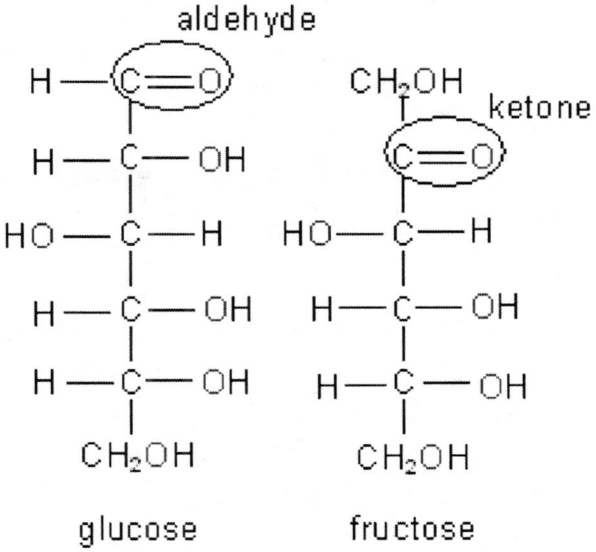

glucose   fructose

Glucose and fructose are the same in number of C, H, and O. One can see that aldehyde is the same as ketone in composition, but the location in glucose is different from the fructose.

Note that glucose and other kinds of sugars may be linear molecules or as in aqueous solution they become a ring form (hexagonal or pentagonal form).

Different shapes of molecules that are composed of the same number and kinds of atoms are called isomers. Glucose and fructose (shown below) are both $C_6H_{12}O6$ but the atoms are arranged in different configuration in each molecule, Figure (15).

Figure (15): Isomers of glucose

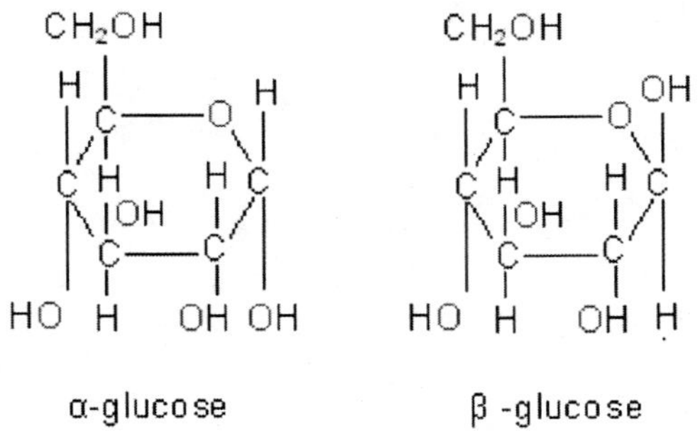

α-glucose   β-glucose

There are two isomers of the ring form of glucose. They differ in the location of the OH and H groups.

2-7-2-2 Polysaccharides

Polysaccharides are a chain of Monosaccharide molecules. Here are some types of polysaccharides:

2-7-2-3 Starch and glycogen

The function of starch and glycogen, Figure (16), is to store energy. They are composed of glucose monosaccharide (monomers) bonded together producing long chains.

Figure (16): Glucose or starch

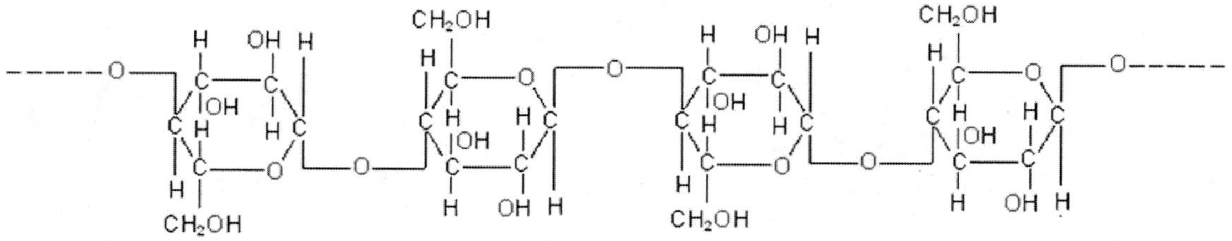

Glycogen is stored in the liver and muscles. During fasting or between meal times, the liver releases glycogen (by an enzyme called glucagon) in order to balance the sugar (glucose). Extra glucose is stored back in the liver as glycogen.

Plants produce starch to store carbohydrates and converted back to energy through the photosynthetic process.

2-7-2-4 Cellulose and chitin polysaccharides

Cellulose supports and protects the cell walls of plants. The cell walls of fungi and the exoskeleton of arthropods are composed of chitin, Figure (17). Cellulose is shown in Figure (18).

Figure (17): Polysaccharides of chitin

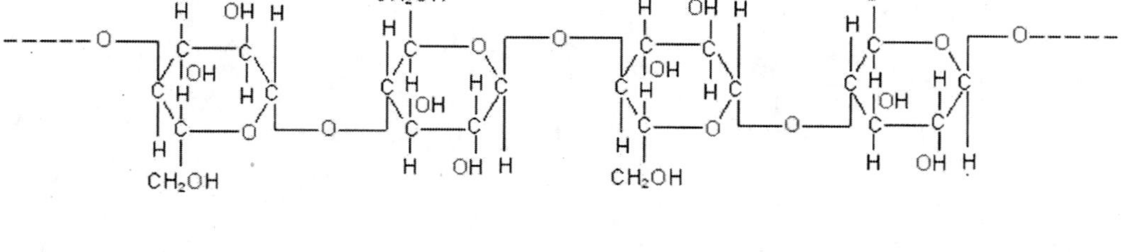

Figure (18): Polysaccharide of cellulose

Humans and most animals do not have the necessary enzymes required to digest the cellulose or chitin. Animals often have microorganisms in their guts that convert them to sugar. Fiber is cellulose which is an important component of the human diet.

With energy from light (photosynthesis), plants can build sugars from carbon dioxide and water. Glycerin (also called glycerol) is not a sugar but is basically one half of a glucose sugar.

Humans and most animals do not have the necessary enzymes needed to digest the cellulose or chitin. Animals that can digest cellulose often have microorganisms in their gut that digest it for them. Fiber is cellulose, an important component of the human diet.

With energy from light, plants can build sugars from carbon dioxide and water. Glycerin (also called glycerol) is not a sugar but is basically one half of a glucose sugar, Figure (19).

Figure (19): Glucose and Glycerin

2-7-2-5 Nucleotide; sugar, nitrogenous hydroxyl, and phosphate

DNA and RNA are composed of nucleotides. Nucleotides consist of three joined structures: a nitrogenous hydroxyl, a sugar, and a phosphate group. The joined sugar is either ribose or deoxyribose. Nucleotides are the main structure of DNA and RNA which are important biopolymers in cellular genetic coding and metabolism. DNA has four nucleotides as does RNA of which three nucleotides are common with those of the DNA. The difference is that DNA has Thymine, and RNA has Uracil nucleotide, Figure (20).

Figure (20): Two stranded of the DNA with its components; nucleotides, phosphate, and sugar connections

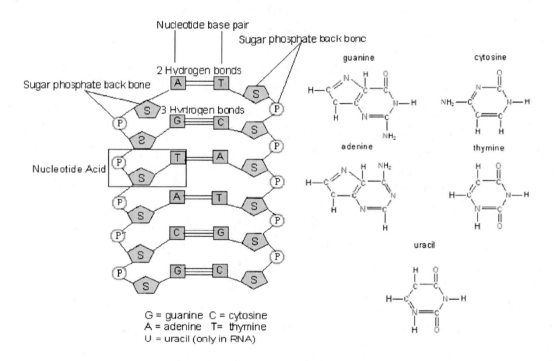

Monosaccharides are classified by the number of carbon atoms as below:

- Triose   3 carbons atoms

- Tetrose   4 carbon atoms

- Pentose   5 carbon atoms

- Hexose   6 carbon atoms

- Heptose   7 carbon atoms

- Octose   8 carbon atoms

- Nonose   9 carbon atoms

- Decose   10 carbon atoms

When monosaccharide ends with -CHO, it is aldehyde, and ketose when it ends with C=O.

2-7-3 Lipids

Study showed that cholesterol/ phospholipid (C/PL) molar ratio significantly increased with aging. This would play a role in the onset of atherosclerosis and in thrombotic phenomena occurring with increasing frequency in the elderly. Because lipids are a major component of living organisms and probably the first easy target of free radicals once they are produced, lipid peroxidation might play an important role in initiating and/or mediating some aspects of the aging process. It has been widely demonstrated that there is an age-associated increase in the steady-state concentrations of lipid peroxidation products.

The lipids of cellular membranes not only serve roles in controlling the structure and fluidity of the membrane, but are increasingly recognized for their roles as signalling molecules and modifiers of membrane protein function. Recent studies reveal striking changes in membrane lipids during aging and in age-related diseases such as cancer, cardiovascular disease and neurodegenerative disorders. Lipids including inositol phospholipids, cholesterol, sphingolipids, choline, and ceramides play important roles in signalling cellular responses to stress and specific stimuli such as growth factors, cytokines and neurotransmitters.

Lipids have three functions; energy storage, forming the membrane around our cells, and hormones and vitamins.

Each type of lipid has a different structure and they have a large number of carbon and hydrogen bonds which makes them non lone polar group of molecule. Oxygen which is stronger than carbon in separating and pulling the hydrogen from carbon makes lipids very energy-rich.

As we mentioned before that water has two lone poles, the atoms are too strong to be separated from each other, and accordingly they do not attached to the carbon atoms in the lipid. Thus lipids are insoluble in the water, and they are stored in our body which has a large amount of water.

Most lipids are composed of some sort of fatty acid arrangement. The fatty acids are composed of methylene (or Methyl) groups.

138

Fatty Acids: Acid means (−OH). Unsaturated fatty acid is a chain of Methylene with at least two carbon atoms lost their hydrogen atoms, as shown in Figure (21).

Figure (21): Unsaturated and saturated fatty acid

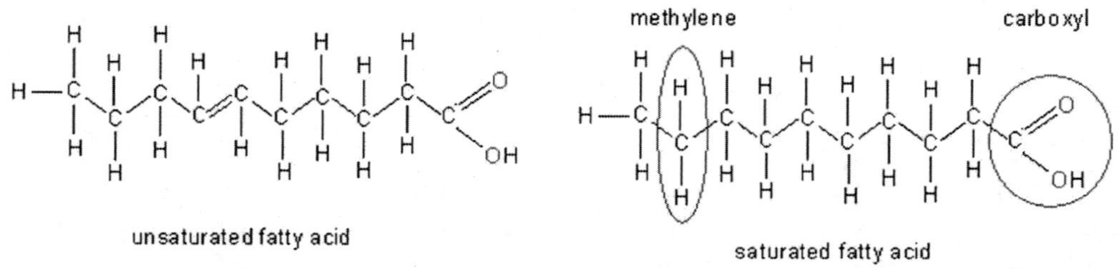

unsaturated fatty acid

saturated fatty acid

The fatty acid chains are usually between 10 and 20 Carbon atoms long. The fatty "tail" is non-polar (Hydrophobic; hates water) while the Carboxyl "head" is a little polar (Hydrophilic, loves water).

A fat is a solid at room temperature, while oil is a liquid under the same conditions. The fatty acids in oils are mostly unsaturated because they have less hydrogen which is the lightest in the periodic table, while those in fats are mostly saturated, and therefore float such as butter and margarine.

The double bond also gives unsaturated fatty acids a strong bond (denser) in the methylene chain. And stick to each other. These interactions make them less fluid and more solid. Figure (22) shows physical chains of saturated and unsaturated fatty acids.

Figure (22): Pictures of fatty acids (physical shape)

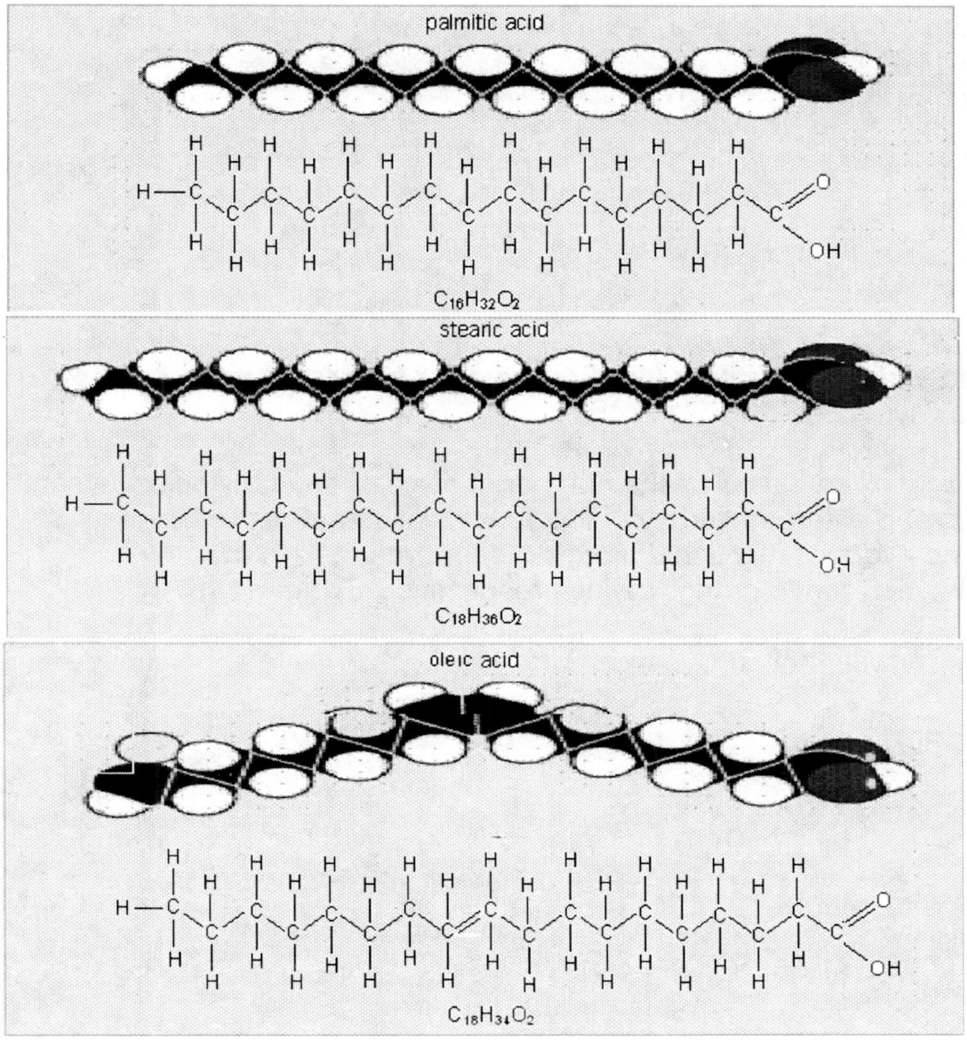

palmitic acid

$C_{16}H_{32}O_2$

stearic acid

$C_{18}H_{36}O_2$

oleic acid

$C_{18}H_{34}O_2$

Animals convert excess sugar to glycogen that is stored in the liver. Excess glycogen will be converted into fat. Most plants convert excess sugar into starch. Some seeds and fruits store energy as oils (canola and sunflower). Fat yields 9.3Kcal/gm, and carbohydrate yields 3.79 Kcal/gm. Fat thus store energy as 6 times as sugar.

The human body stores some fat under the skin in the sub dermal layers as insulation to protect him from tough environment.

2-7-3-1 Types of lipids

Fatty Acids
- Saturated
- Unsaturated
Glycerides

- Neutral
- Phosphoglycerides
Complex Lipids
- Lipoproteins
- Glycolipids
Nonglycerides
- Sphingolipids
- Steroids
- Waxes

Since saturated and unsaturated fatty acids have been discussed, we can turn our attention to Glycerides.

Glycerides are classified based on the number of glycerol and fatty acid. Glycerol is a functional group of three hydroxyls. The fatty acids can react with one, two, or all three of the hydroxyl functional groups of the glycerol to form monoglycerides, diglyceride or triglycerides respectively, Figure (23).

Figure (23): Monoglycerides, diglycerides, triglycerides

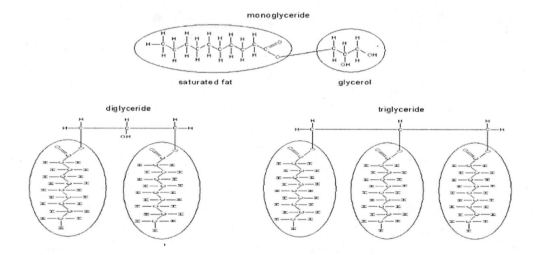

Triglycerides, as main components of very low density lipoprotein (VLDL) play an important role in metabolism as energy sources and transporters of dietary fat. In the intestine, triglycerides are split into glycerol and fatty acids (this process is called lipolysis), (through a process called lipolysis), with the help of lipases and bile secretions, which are then transported into the cells lining of the intestine, and absorbed by the absorptive enterocytes.

The triglycerides are rebuilt in and absorbed by the enterocytes from their fragments and packaged together with cholesterol and proteins to form chylomicrons. These are excreted from the cells and collected by the lymph system and transported to the large vessels near the heart before being mixed into the blood. Various tissues can capture the chylomicrons, releasing the triglycerides to be used again as a source of energy. Liver cells can synthesize and store triglycerides.

When the body requires fatty acids as an energy source, the hormone glucagon signals the breakdown of the triglycerides by hormone-sensitive lipase to release free fatty acids. As the brain can not utilize fatty acids as an energy source, the glycerol component of triglycerides can be converted into glucose for brain fuel when it is broken down, Figure (24). Fat cells may also be broken down to feed the brain, if the brain, if needed.

Figure (24): Conversion of triglyceride into chylomicrons, fatty acid and glycerol

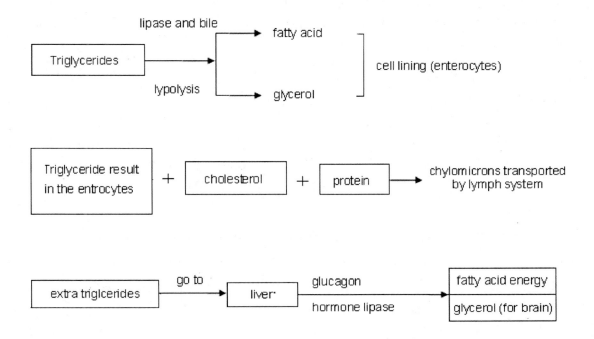

High density lipoprotein (HDL) is the most helpful in preventing coronary heart disease.

2-7-4 Proteins

A majority of our muscles and organs are made of protein. Muscle mass is critical for performance of the activities of daily living. Although the recommended dietary allowance is currently the same for all adults, 0.8g complete protein/kg body weight, there is considerable evidence that the protein requirement is probably greater for older adults. It is possible that the protein requirement for healthy older adults is as high as 1.0g or even 1.2g/kg body weight.

Our body uses 20 amino acids to make body proteins. There is a dietary requirement for nine indispensable amino acids (also known as essential amino acids) that cannot be made by the body. Proteins are considered to be of high quality if they provide all of the indispensable amino acids. The body also requires an additional increment of protein every day that can be used to make the dispensable amino acids in the proportions that are needed. The protein, Klotho, is found in several species. In mice, the researchers discovered, it acts as a hormone, circulating through the blood and binding to cells. Therapies based on this hormone could prove to be a way to extend life or slow its effects.

Amino acids contain both a carboxyl group (COOH) and an amino group (NH_2). The general formula for an amino acid is given below in Figure (25) below. Amino acids are either polar or non polar charges. The acidic positive COOH (COO^-) and the negative NH3 ($NH3^+$) cancel each other in some amino acids. Some amino acids molecules have larger charge in one side than the other and polar bonds.

Figure (25): Polar and non-polar amino acid

$$NH2$$
$$R— C —COOH$$
$$H$$

polar

R is the functional group of the amino acid, and can be replaced by H

$$NH_3^+$$
$$R— C —COO^-$$
$$H$$

non-polar

Proteins have 20 amino acids and have different combinations of chains of amino acids.

2-7-4-1 Non-Polar side chains

There are eight amino acids with non-polar side chains. Glycine, alanine, and proline have small, non-polar side chains and are all weakly hydrophobic. Hydrophobic molecule is not dissolved in water, as hydrophobia means something with a fear of water. Hydrophobic molecules often group together when dropped in water which is polar side chain, just as oil and butter do. Hydrophilic molecules can dissolve in water. Fattyphilic or lipophilic (philic means love) can also dissolve in oils and other lipids. They tend to be electrically neutral and non-polar and work better with neutral and non-polar solvents.

2-7-4-2 Polar; uncharged side chains

There are also eight amino acids with polar, uncharged side chains. Serine and threonine have hydroxyl groups. They are hydrophilic and have distinguished smell. Histidine and tryptophan have heterocyclic aromatic amine side chains. Cysteine has a sulfhydryl group. Tyrosine has a phenolic side chain. The Twenty amino acids are shown in Table (3).

Table (3): Amino Acids (20 amino acids)

Name	Abbreviation	Linear Structure
Alanine	ala A	CH3-CH(NH2)-COOH
Arginine	arg R	HN=C(NH2)-NH-(CH2)3-CH(NH2)-COOH
Asparagine	asn N	H2N-CO-CH2-CH(NH2)-COOH
Aspartic Acid	asp D	HOOC-CH2-CH(NH2)-COOH
Cysteine	cys C	HS-CH2-CH(NH2)-COOH
Glutamic Acid	glu E	HOOC-(CH2)2-CH(NH2)-COOH
Glutamine	gln Q	H2N-CO-(CH2)2-CH(NH2)-COOH
Glycine	gly G	NH2-CH2-COOH
Histidine	his H	NH-CH=N-CH=C-CH2-CH(NH2)-COOH
Isoleucine	ile I	CH3-CH2-CH(CH3)-CH(NH2)-COOH
Leucine	leu L	(CH3)2-CH-CH2-CH(NH2)-COOH
Lysine	lys K	H2N-(CH2)4-CH(NH2)-COOH
Methionine	met M	CH3-S-(CH2)2-CH(NH2)-COOH
Phenylalanine	phe F	Ph-CH2-CH(NH2)-COOH
Proline	pro P	NH-(CH2)3-CH-COOH
Serine	ser S	HO-CH2-CH(NH2)-COOH
Threonine	thr T	CH3-CH(OH)-CH(NH2)-COOH
Tryptophan	trp W	Ph-NH-CH=C-CH2-CH(NH2)-COOH
Tyrosine	tyr Y	HO-Ph-CH2-CH(NH2)-COOH
Valine	val V	(CH3)2-CH-CH(NH2)-COOH

Brain is one of the organs rich in phospholipids, which provide the building blocks for different membrane structures. These phospholipids are rather rich in long-chain polyunsaturated fatty acids, particularly in docosahexaenoic acid and arachidonic acid. Phospholipids and their nonpolar side chains not only are essential constituents of all cellular membranes but they can affect the functional activities of the membranes.

Phospholglycerides are esters of only two fatty acids, phosphoric acid and a trifunctional alcohol (glycerol) as revealed in Figure (26). The fatty acids are attached to the glycerol at two positions on glycerol through ester bonds. There may be a variety of fatty acids, both saturated and unsaturated, in the phospholipids.

The third oxygen on glycerol is bonded to phosphoric acid through a phosphate ester bond. In addition, there is usually a complex amino alcohol also attached to the phosphate through a second phosphate ester bond. The complex amino alcohols include choline, ethanolamine, and the amino acid-serine.

The properties of phospholipids are characterized by the properties of the fatty acid chain and the phosphate/amino alcohol. The long hydrocarbon chains of the fatty acids are of course non-polar. The phosphate group has negatively charged oxygen smaller than the positively charged nitrogen to make this group ionic. Note that the oxygen bonds with the carbon which gives one electron. The charge between them is weak because the attraction force is proportional to the multiplication of 4 and 6 of the external orbits of the carbon and the oxygen atoms, and the result is weak positive oxygen as the oxygen pulls one electron from the carbon (oxygen is more negative than carbon and wants to finish its external orbit to be 8 electron).

Phospholipids are major components in the lipid bilayers of cell membranes.

Figure (26): Phosphoglyceride

diglycerides   glycerol   phosphate   amino acid

phosphoglyceride

Table (4) Classification of lipids

Type of fatty acid	
Fatty Acids	Glycerides
Saturated Fatty Acids	Triglycerides
Unsaturated Fatty Acids	Phosphoglycerides
Soap (salt of fatty acid)	
Prostaglandins	
Non glyceride Lipids	
Waxes	Steroids
Sphingolipids	Lipoproteins

Table (5) Types of simple lipids

Birth control pills	Synthetic Steroid
Anabolic Steroids	Testosterone, not allowed in sport
Olestra	Fat substitute without calories
Detergents and Surfactants	Soaps and cleaners
Micelle	For water separation
Hydrogenation	Converting unsaturated to saturated as in margarine

2-7-4-3 Examples of long chain fatty acids

- Arachidic acid ($C_{20}H_{40}O_2$), also called eicosanoic acid, is a saturated fatty acid found in peanuts oil. Fatty acids are a carboxylic acid with a long unbranched aliphatic tail, which is either saturated or unsaturated. It can be formed by the hydrogenation of arachidonic acid. It is practically insoluble in water, and stable under normal conditions, Figure (27).

Figure (27): Arachidic acid

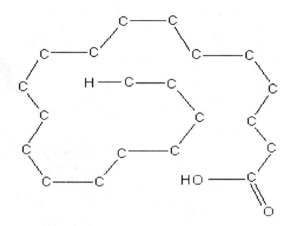

Each carbon atom is connected to
two hydrogen atoms

- Arachidonic acid

There are two types of fats that fall into the category of "healthy" fats. These are the monounsaturated fats and the long chain omega 3 and omega 6 fats. Monounsaturated fats are found in olive oil, some nuts and avocados. Long chain omega 3 and omega 6 fats come from fish, fish oils, and eggs. These are exceptionally powerful nutrients in your quest for a longer and healthier life.

This polyunsaturated arachidonic fat, Figure (28), may be the most dangerous fat known when consumed in excess and is known as an Omega 6 fat. If you inject virtually every type of fat, saturated fat and unsaturated, into rabbits, nothing will happen. However, if you inject an arachidonic acid into the same rabbits they are dead within few minutes. The human body needs "some" arachidonic acid, but too much of it can be toxic.

Figure (28): Arachidonic acid

o Eicosatrienoic acid

Eicosatrienoic acid, Figure (29), is rich in omega 3. Researchers have discovered that fat is essential for good health. They've also determined that all fats are good. But if the balance of fats is unequal, then that leads to body health problems within the body. The balance of fats (the ratio of Omega-6 fatty acids to Omega-3 Fatty Acids) in the membranes of the cells in our bodies should be 1 : 1. When the ratio is changed, for instance when it's greater than four to one, body problems start to occur.

Figure (29): Eicosatrienoic acid

2-8 Esters

The product of a condensation reaction in which a molecule of an acid unites with a molecule of alcohol with elimination of a molecule of water as shown in the following equation, Figure (30).

Figure (30): Condensation of acid and alcohol

acid   alcohol   ester   water

148

Some other types of esters are shown in Figure (30-a).

Figure (30-a): Other types of esters

| carboxylic acid ester | nitric acid ester | phosphoric acid ester | sulfuric acid ester |

Esters can be represented in many forms. Figure (31) shows esters in the form of methyl propanoate (3 carbons), ethyl ethanoate, ethanoyl chloride, propanamide, and hydroxypropanentrite acids

Figure (31): Different Forms of esters

$CH_3 - CH_2 - C$ (with =O and O—H)	Propanoate combines with methyl (CH3) which is alkyl group.	Methyl propanoate acid
$CH_3 - CH_2 - C$ (with =O and O—CH₃)	This stronger methyl propanoate replaces H with additional methyl group.	Methyl propanoate acid
$CH_3 - C$ (with =O and O—CH₂—CH₃)	Ethyl group is C2H5 or CH2-CH3. Methyl is connected to ethyl group through COOH, and the acid has the characteristic of ethylene	Ethyl ethanoate acid
$—C$ (=O, Cl) $CH_3 — C$ (=O, Cl)	The one on the left is acyl chloride, and the one on the right is ethnoyl chloride.	Ethanoyl chloride acid
$CH_3 — CH_2 — C$ (=O, NH₂)	Propane and ester amino group	Propanamide acid
$CH_3 — CH — C ≡ N$ (with OH)	Ol, ide, and ate means 1, 2, 3 or more oxygens. This is called hydroxypropanentrite and not hydroxypropanentole. See IUPAC nomenclature later in this chapter.	Hydroxypropanentrite acid

We haven't finished with lipid yet, lipids are classified into two subgroups; the simple group and the complex group.

  o Simple lipids

     Simple lipids contain C, H, and O atoms. They can be separated into four types:

  a- Fatty acid (FA) is usually a long chain of monocarboxylic acids. Fatty acids with small chain of carbon and hydrogen (4 - 6) can easily attract oxygen and become oxidized. Oxidized fatty acids are thicker and waxy and can be converted into sterol and cholesterol.
  b- Waxy lipids have long hydrocarbon tails on polar head (double bonded oxygen). Paraffin is 100% carbon and hydrogen chains (like lipid tails) and at room temperature they are about 9 carbons up, they are solid at room temp and are waxy. The longer the chain, the harder and dryer the wax is.
  c- Triglycerides have been discussed in previous sections.
  d- Sterols is any of a group of naturally occurring or synthetic organic compounds with a ring structure of steroid (steroid has three fused cyclohexane plus a fourth cyclopentane, see Figure (32), having an OH (hydroxyl) group, usually attached to carbon-3. This hydroxyl group is often connected with a fatty acid by esterfication (or without esterification) such as the cholesterol as shown in Figure (32).

Figure (32): Cholesterol and steroid

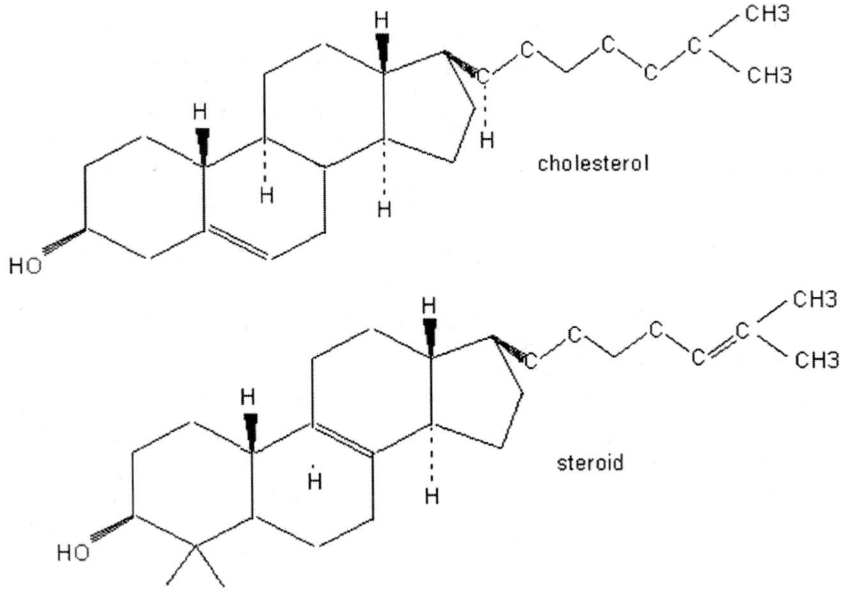

Sterol group is one of the cholesterol groups. Cholesterol and steroid (steroid will be discussed later) have similar molecular formulas. When two steroids join together, water will come out of the bond, and the cholesterol is now non-polar and can not dissolve in water.

Cholesterol is a fatty substance produced mainly by the liver, and stored in the entire body; It is a vital part of the whole body in order for it to function appropriately. Cholesterol helps make vitamin D and some hormones in the body. The human body makes about 80% of our cholesterol while the other 20% comes from our diet. The main causes of high cholesterol are heredity, fatty diet and being overweight.

Food cholesterol comes mainly from animal products: meat and liver, eggs, milk products, butter, etc. The type of food (saturated and hydrogenated fats) affects blood cholesterol. Therefore, you need to know which diet contains saturated and hydrogenated fats if you want to reduce your blood cholesterol levels.

Cholesterol doesn't mix easily with blood as explained before. There are two types of carriers: LDLs (Low-Density Lipoprotein), known as "bad cholesterol" and HDLs (High-Density Lipoprotein), known as "good cholesterol." When cholesterol is carried by the blood, LDLs tends to stick to artery walls, forming deposits.

These deposits can cause a heart attack or stroke when they partially or completely obstruct an artery and block the blood flow. HDLs travel in the blood and from the artery walls to the liver, where bile gets rid of it.

- o Complex lipids

  Complex lipids have additional components to the simple lipids such as phosphoric acids. Among the complex lipids, are phosphoglycerides, phosphosphingolipids, and glycolipids. The phosphoglyceride and phosphatidic acid is similar in structure to a triglyceride except that the phosphoric acid is esterified to the 3 hydroxyl groups of the glycerol component rather than to the FA, Figure (33).

Figure (33): Phosphoglyceride molecule attached to other molecules

phosphoglyceride

Esterification of the phosphoric acid (PA) and glyceride together with the hydroxyl groups lead to a series of phosphoglycerides, including phosphatidyl choline (PC), commonly known as lecithin, phosphatidyl ethanolamine (PE), and phosphatidyl serine (PS). Acylglycerol and diacylglycerol phosphate, Figure (34) are constituents of nerve tissues and involved in fat storage and transport.

Figure (34): Acylglycerol and diacylglycerol phosphate from glycerol phosphate

Cholesterol is a fatty substance produced mainly by the liver and go to the proper functioning of the entire body. Cholesterol helps make vitamin D and some hormones in the body. Our bodies make about 80% of our cholesterol; the other 20% comes from our diet. The main causes of high cholesterol are fatty diet and, being overweight and heredity.

2-9 Lipoproteins

Lipoproteins are molecules that have fat and protein. They carry all kinds of cholesterol and other similar substances through the blood. Lipoproteins are the essential component of proteins, antigens, transporters, adhesion, and toxins that will be discussed in chapter 2 outlining biology. High level of lipoproteins can increase the risk of heart disease which results in an atherosclerosis stroke and myocardial infarction (heart attack). Figure (33) above shows one molecule of phosphoglyceride (lipoprotein) of which two stearic acids (nonpolar) and phosphatidylcholine (phospholipid) which has polar head are connected together. As explained previously that the polar is hydrophilic and is soluble in water, where as the fat is insoluble. The lipoprotein molecule of this structure spontaneously form aggregate structures such as micelles and lipid bilayers with their head oriented toward watery medium and their tail shielded from the water, Figure (35).

Figure (35): Lipoprotein (phosphatidylcholine bilayer) with polar head and non-polar trails

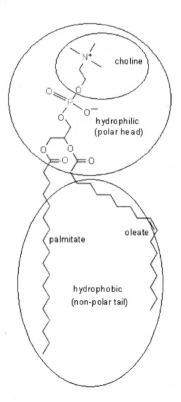

2-10 Glycolipids

Glycolipids are lipids that are attached to a short carbohydrate chain (mono- or oligosaccharides which is saccharide polymer containing a small number of carbon atom, oligo means 'few' in Greek). The role of the glycolipids is to serve as transporters from cell to cell and to provide energy. Glycolipids are collectively part of a larger family of substances known as glycoconjugates (carbohydrates covalently bonded with other chemical species), which include glycoproteins, glycopeptides, glycolipids, proteoglycans, peptidoglycans and lipopolysaccharides which all the same except the way of bonding. Glycolipids are collectively part of a larger family of substances known as glycoconjugates. Glycoconjugates are carbohydrates covalently bonded with other chemical species which include glycoproteins, glycopeptides, glycolipids, proteoglycans, peptidoglycans and lipopolysaccharides. They are the same except for the way they bond.

Here are some molecules of glycopeptides with different names, Figure (36).

Glycolipids are the main component of the outer surface of cell membrane and linked with phospholipids in the cell surface. Protein link-lipid monosaccharides are a large component of the eukaryotic cell surface (eukaryotic cell has a nucleus, where prokaryotic cell has no nucleus) where they play an important roles in cell-pathogen (germ) interaction and adhesion.

Mutation or catabolism in the link of glycolipids and phospholipids could have serious threatening consequences due to the oxidation of lipids molecules. Oxidation will be discussed in detail in this chapter. This could lead to some types of tumour metastasis. Oxidation could also cause sugar deficiency and lisosomal storage disease which affects the liver glycogen metabolism. Mucin is one important glycolipid for the immune system and other connecting factors in bones and gonad system.

Figure (36): Galactocerebroside and canglioside as glycopeptides

154

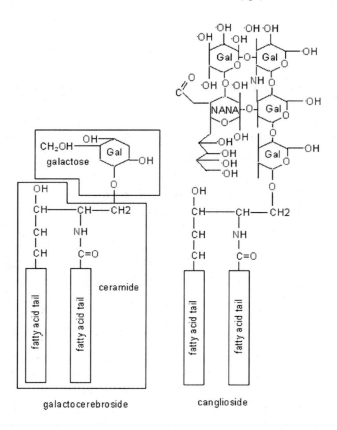

galactocerebroside   canglioside

2-11 Sphingolipids

Sphingolipids are a class of lipoprotein found in neural tissues, and play an important role in cell recognition and signal transmission. There are mainly four lipoproteins, Figure (37), which play an important role in cell recognition and signal transmission:

- Sphingomyelin is amide link (acyl link) and methyl group. The amid link can be linked to amino acids in the RNA and helps in communication, RNA function is further discussed in the second chapter.
- Phosphtidylcholine which has the same molecule as sphingomyelin except that ester is the link.
- Phosphatidylserine has a charged negative carboxyl ion
- Phosphatidylethanolamine is similar to phosphatidylserine but neutral and has no carboxyl.

Figure (37): Four types of liporoteins

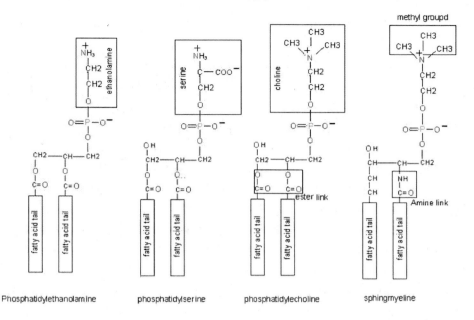

Phosphatidylethanolamine   phosphatidylserine   phosphatidylecholine   sphingmyeline

2-12 Steroids

A steroid is a chain of three hexa-rings and one penta-ring of carbon. Steroids vary by the functional group that are attached to these rings. Functional groups of O and OH and their location can create hundreds of steroids. Figure (38) shows different steroids molecules of different function in human. Steroids are also made in plants and fungi. Estrogen and progesterone are made in the ovary and placenta before and after the 15 days of the menstrual period respectively Testosterone is made in the testes, and some of it is also made in female to control the estrogen. Cholesterol in the mitochondrion, as explained in second chapter, is converted to the required steroid. Estrogen is the main component of contraceptive and hormone replacement therapy (HRT). Estrogen has OH at the end, and androgen has O at the end of the steroid molecule. Both men and women have both estrogen and androgen. In male, androgen is synthesized rapidly and same thing occurs for the estrogen in female. After menopause, the estrogen is gradually reduced and the androgen is emerged to produce hair in older women. Some estrogens are also produce in the liver, adrenal gland and the breasts to control the mode of postmenopausal women. HRT is given to postmenopausal women to prevent osteoporosis. Estrogen controls the HDL and LDL cholesterol. Androgen controls the sex desire in both male and female. Hormone replacement therapy (HRT) should not be given after the start of pregnancy, because the level of progesterone is higher than the estrogen, otherwise the risk of cancer could be positive. Once the cancer established in the breast, the cancer loves the estrogen hormone. These cancers can be treated by suppressing the estrogen.

Figure (38): Androgen and estrogen steroids

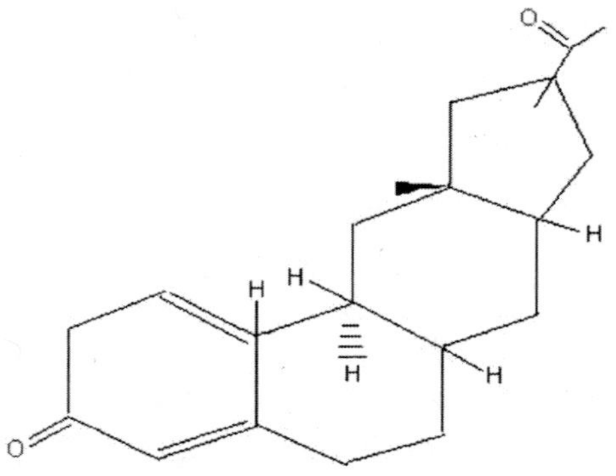

androgen/progesterone

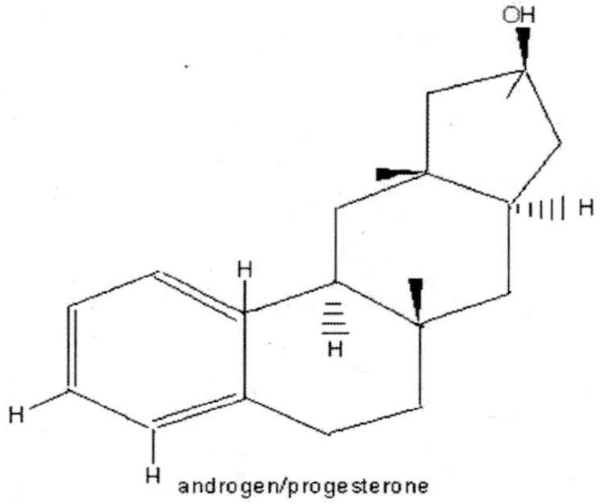

androgen/progesterone

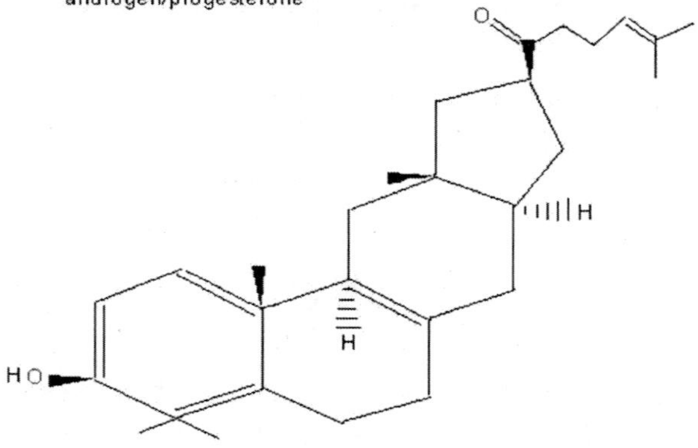

steroid

2-13 Mucin

Mucin is one example of glycoproteins found in the mucous of the lung. Mucins are hydrophilic and accordingly resist the process of proteolysis due to the deficiency of water as mucins absorb water. Sugar with double bond oxygen absorbs the water in the mucine, and drive out water in the protein and cause mucosal barrier which may be important in maintaining mucosal barriers mucins, and then the water is secreted in the mucus of the respiratory and digestive tracts. The sugars attached to mucins give them considerable water-holding capacity and also make them resistant to the forming of proteins. Mucin is important for the white blood cell to fight and kill proteins of germs and viruses mucins are a main building block of the immune system. Examples of glycoproteins (mucins) are:

- Immunoglubin (antibodies) which interact with antigens (germ).
- Molecules of major histocompatibility complex (MHC) which is on the surface of the cell and interacts with T cells. Refer to T cells and MHC in the second chapter.
- Platelets in which mucins drive out the water in blood coagulation.
- Sperm-egg attraction is caused by the interaction between the =O and -OH in both the sperm and the egg in the zona pellucida.
- Connective tissue in bones and skeleton.

- Glycoprotein (mucin) also enters in the hormonlysis such as follicle-simulating hormone, thyroid simulating hormone, chorionic gonadotropin, Luteinizing hormone, alpha-fetoprotein, and erythropoietin.

Mucin is a mixture of glucose and protein. Three different molecule structures of mucin are shown in Figure (39).

Figure (39): Configuration of three types of mucin

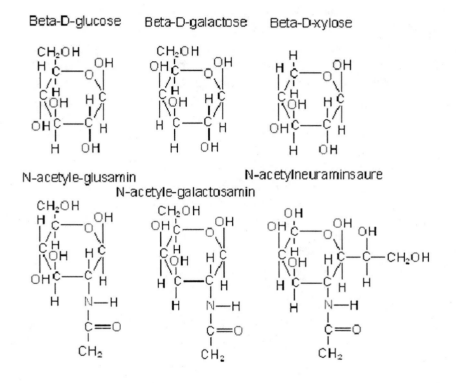

2-14 Oxidation and reduction

Enzymes in the human body regulate oxidation-reduction reactions. These complex proteins, of which several hundred are known, act as catalysts, speeding up chemical processes in the body. Oxidation-reduction reactions also take place in the metabolism of food for energy, with substances in the food broken down into components the body can use.

Weindruch, who is studying calorically restricted monkeys at the University of Wisconsin- Madison. Writing in Scientific American, wrote that mitochondria seem to inadvertently make free radicals while producing ATP. The free radicals, in turn, damage the mitochondria, and the damaged mitochondria make more free radicals. Eventually the cycle of damage becomes a slippery slope that we call "aging", because all amino acid residues of proteins are susceptible to oxidation by the free radicals. Oxidation of the protein backbone results in protein fragmentation.

Oxidation is defined as the addition of oxygen, and the reduction is the removal of oxygen. Oxidation and reduction can also be defined as the loss and gain of electrons respectively. The loss of electrons from an atom produces a positive oxidation state, while the gain of electrons results in a negative oxidation state.

From the periodic table, one can simply observe that oxidation is associated with metals, and reduction is associated with non-metals. All metal atoms are characterized by their tendency to lose one or more electrons, forming a positively charged ion (oxidation), called a cation. During this oxidation reaction, the oxidation state of the metal always increases from zero to a positive number, such as "+1, +2, +3, etc. This depends on the number of electrons lost from the metal or gained by the non-metal. When the non-metal gains a negative charge (electrons), it is called anion. See Table (6).

Table (6): Group number and charges

Group Number	Number of electron lost or gained (L $ G)	Charge cation or anion (C & A)
I Hydrogen	1 L	+1 C
II Beryllium	2 L	+2 C
III Boron	3 L	+3 C
IV Carbon	4 L or 4 G	+4 C or - 4 C
V Nitrogen	3 G	-3 A
VI Oxygen	2 G	-2 A
VII Fluorine	1 G	-1 A
VIII Neon	0 Stable	0 Neutral

The number of electrons in the outermost orbit is often called valence electrons. Electro negativity in metals is lower than in non-metals. It is therefore that metal atoms lose electrons to non-metals atoms. Table (5) shows negativity in some metal and nonmetal atoms.

Table (7): Negativity of some metals and nonmetals.

Metallic elements			Nonmetallic elements			
Li	Be		C	N	O	F
1	1.5		2.5	3	2.5	4
Na	Mg	Al		P	S	Cl
1	1.2	1.5		2.1	2.5	3

K	Ca	Sc	Fe	Ni	Cu	Zn		Se	Br
---	----	----	----	----	----	----		----	----
0.9	1	1.3	1.8	1.9	1.9	1.7		2.5	2.8

By convention oxidation and reduction reactions are written in the following form: as an example:

By convention reduction reactions are written in the following way:

Atom		Number of electrons		Atom charge
X	+	n (e^-)		X^{-n}
X	-	n (e-)		X^{+n}

The carbon for example has two states:

C	+	$4e^-$		C^{-4}
C	+	4 (-e)		C^{+4}

As the non-metal atom gains the electrons lost by the metal, it reduces its state of negativity from zero to a negative value (-1,-2.-3) depending on the number of electrons gained by the non-metal, Figure (40).

Figure (40): Oxidation and reduction in metal and nonmetal

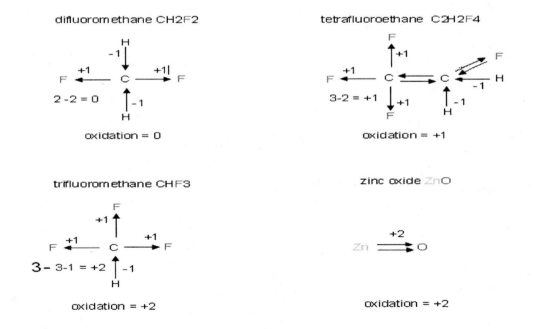

Note that the GROUP VIII non-metals have no tendency to gain additional electrons; hence they are non-reactive in terms of oxidation-reduction. This is one the reasons why this family of elements was originally called the inert gases.

Oxidation-reduction reactions mean that the process of oxidation cannot occur without a corresponding reduction reaction. Oxidation must always be "coupled" with reduction, and the electrons that are "lost" by one substance must always be "gained" by another, Figure (41).

Figure (41): Oxidation of some molecules

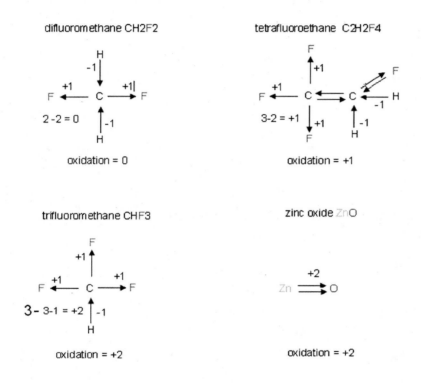

2-14-1 Oxidation and reduction in terms of oxygen

- Oxidation is gain of oxygen
- Reduction is loss of oxygen

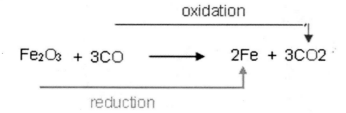

The iron oxide is called oxidizing agent, and the carbon monoxide is called reducing agent. In other word.

Reduction and oxidation are known as a redox reaction.

2-14-2 Oxidation and reduction in terms of hydrogen transfer

- o Oxidation is loss of hydrogen.
- o Reduction is gain of hydrogen.

Oxidation and Reduction are illustrated bellow:

oxidation by loss of hydrogen

$CH_3CH_2OH \longrightarrow CH_3CHO$

Ethanol　　　　　　　acetaldehyde

reduction by gain of hydrogen

$CH_3CHO \longrightarrow CH_3CH_2OH$

acetaldehyde　　　　　Ethanol

- o Oxidizing agents give oxygen to another substance.
- o Reducing agents remove oxygen from another substance

$$\overset{0}{Cu} + 2Ag\overset{+1}{N}\overset{+5}{O_3}\overset{-2}{} \longrightarrow \overset{+2}{Cu}(\overset{+5}{N}\overset{-2}{O_3})_2 + 2\overset{0}{Ag}$$

2 x (+1+ 5 -2)　　　2 + (5 -2)₂
2 x 4 = 8　　　　　2 + 6 = 8

Note: O3 is considered as one oxygen, see the diagram below:

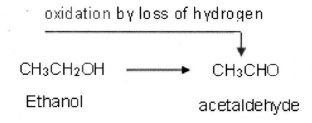

2AgNO3　　　　　　　Cu (NO3)₂

Copper in the above equation (left hand side) as has zero charge, as it stands alone. The Oxygen in AgNO3 is the most negative atom, and therefore it tends to take all electrons from Ag and N. Therefore, the Oxygen will have (-2) charge, N will have +5, and g will have (+1).

The left side of the equation will have Cu (+2), N (+5), and O (-2). If we equate both sides of the equation, then both sides are equal to zero charge.

2-14-3 Examples

2-14-3-1 Sulphur Trioxide

$$2S + 3O2 \longrightarrow 2SO3$$
$$0 \quad\quad 0 \quad\quad\quad +6\ -2$$

Note that both oxygen and sulfur ends with 6 electrons in the outermost orbit, but the oxygen is more negative, because the second orbit in oxygen is 6 and the third orbit in sulfur is also 6.

2-14-3-2 Water

$$2H_2 + O_2 \longrightarrow 2H_2O$$
$$0 \quad\quad 0 \quad\quad\quad +1\ -2$$

Combination reaction can be reversed. For example, a compound can be decomposed into the components from which it was composed. This type of reaction is called decomposition reaction.

2-14-3-3 Potassium Chloride

$$2KClO_3 \longrightarrow 2KCl + 3O_2$$
$$+1\ +7\ -1 \quad\quad +1\ -1 \quad\quad 0$$

Redecompositioning can not happen in some reactions.

2-14-3-4 Calcium Carbonate

$$CaCO_3 \longrightarrow CaO + CO_2$$
$$+2\ +4\ -2 \quad\quad +2\ -2 \quad\quad +4\ -2$$

Note that O3 has only -2 because the carbon atom gets its eight electrons from calcium (2 electrons), Oxygen (two electrons), and the existing four electrons in its outside orbit, as shown in the molecular formula of $CaCO_3$ below:

$$Ca^{++} \quad \overset{\displaystyle O}{\underset{\displaystyle O^- \qquad O^-}{\overset{\|}{C}}}$$

2-14-3-5 Iron and Hydrochloric Acid

$$2\,Fe + 6HCl \longrightarrow 2FeCl_3 + 3H_2$$
$$\;\;0 \quad\;\; +1\;-1 \qquad\quad +2\;-1 \qquad 0$$

2-14-4 Oxidation states of the elements

Chemical elements tend to have their outermost orbits in stable conditions,i.e. 2, 8, 18, and 32. Some examples are outlined.

2-14-4-1 Beryllium (2, 2)

Beryllium has 2 electrons in its inner orbit and 2 in the outside orbit. It tends to give the outside 2 electrons to those elements which are more negative. It is difficult to gain electrons as the outside orbit is further from the number 8. Thus, the oxidation state is +2.

2-14-4-2 Carbon (2, 4)

The outermost orbit is 4, and therefore it needs 4 electrons to complete the orbit to 8, or it gets rid off 4 electrons to keep 2 electrons in the outermost orbit. Carbon oxidation states can be written as -4, -3, -2, -1, +1, +2, +3, +4 accordingly.

2-14-4-3 Nitrogen (2, 5)

Nitrogen oxidation state can be written as -3, -2, -1 N +1, +2, +3, +4, +5. Table (6) shows oxidation states for more elements.

Table (8): Oxidation states of some elements

Oxygen (2,6)	-2, -1 O +1, +2
Na (2,8,1)	-1 Na +1
Al (2,8,3)	Al +1,+3
Si (2,8,4)	-4,-3,-2,-1 Si +1, +2,+3,+4
Cl (2,8,7)	-1 Cl +1,+2, +3, +4, +5, +6, +7

Ti (2,8,10,2)	-1 Ti +2, +3, +4	Note that 10 electrons is not a stable number, so it can accept any number of electrons less than 5 which is (10 + 2 - 4) = 8
Cr (2,8,13,1)	-2, -1 Cr +1, +2, +3, +4, +5, +6 Note that -2 makes13+1+2 - 8 = 8, and +6 makes 13 + 1 + 6 - 8 = 8
Mn (2,8,13,2)	-3, -2, -1 Mn +1, +2, +3, + 4, +5, +6, +7
Fe (2,8,14,2)	-2, -1 Fe +1, +2, +3, +4, +5, +6
Zn (2,8,18,2)	Zn +2, Zn Can not accept more electrons to keep It orbit stable (18 =2+10+8), and can not give more than 2 electrons.
Mo (2,8,18,13,1)	-2, -1 Mo 1, +2, +3, +4, +5, +6. This metal has very weak outermost orbitals and can easily accept or give electrons to other atoms.
I (2,8,18,18,7)	-4, -3, -2, -1 I +1, +3, +5, +7
At (2,8,18,32,18,7)	-4, -3, -2, -1 At +1, +3, +5, +7
U (2,18,32,21,9,2)	+3, +4, +5, +6

Let's see the oxidation and the reduction of some atoms in Table (9).

Table (9): Oxidation and reduction of most human elements

			-1	H	1				
			-1	Li	1				
			-1	Na	1				
				K	1				
-4	-3	-2	-1	C	1	2	3	4	
-4	-3	-2	-1	Si	1	2	3	4	
	-3	-2	-1	Na	1	2	3	4	5
	-3	-2	-1	P	1	2	3	4	5
		-2	-1	O	1	2			
		-2	-1	Si	1	2			

One can observe from the above table that the nitrogen and the phosphorous have the tendency for oxidation more than the other elements in table (8). Nitrogen and Phosphorous are vital elements in the formation of amino acids and the DNA as shown in Chapter 2.

As a general rule, atoms with less electro negativity tend to be more oxidized than atoms with higher negativity. Figure (42) has the first three rows of the last five elements of the periodic table, namely;

B   C   N   O   F

Al Si P S Cl
Ga Ge As Se Br

Figure (42): Negativity of previous elements

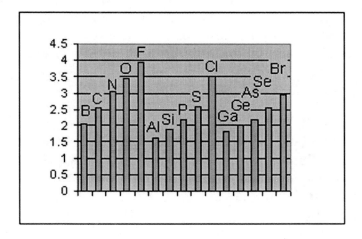

Figure (42): Continued

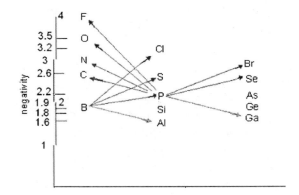

Notice that B (Boron) has less negativity than P (Phosphorous), S (Sulfur), and Cl (Chlorine), and therefore will be oxidized if it is bonded with any of them. However, it will be reduced if it is bonded with aluminum, since the later is lower in negativity. Boron is neither in negativity nor in positivity with Si (Silicon). Similarly, P (phosphorus) is oxidized with respect to C (carbon), N (Nitrogen), O (Oxygen), and F (Fluorine). Ga (Gallium) is oxidized with respect to Phosphorus. The Phosphorus has the tendency for oxidation more than any other element in Figure (42).

Question. If the density of the hydrogen atom is 0.0899 Kg/m^3, and the density of oxygen is 1.429 Kg /m^3, which is larger (in size), a human cell (of same weight) when it is oxidized or when it is reduced? Does that have anything to do with the muscle size? See aerobic and anaerobic respiration in Chapter 2 and Chapter 3.

2-15 Isotopes

Oxidative modifications of cellular components by free radicals are thought to be the cause of ageing and age-associated diseases. Extensive prior research has aimed to lessen such damage by counteracting the free-radical oxidizers with antioxidants. A recent paper describes an original and promising method based on the use of non-radioactive heavy isotopes, which might enable living cells to resist the free-radical oxidation and consequently allow us to live a healthier, longer life.

The strength of a covalent bond is subtly influenced by the atomic masses at either end of the bond: heavier isotopes form stronger bonds than light isotopes of the same elements, and so reactions involving the breaking of those bonds proceed more slowly. For example, hydrogen comes in two stable isotopes, the common 'light' hydrogen isotope (H) and its twice-heavier sibling, deuterium (D). The C–D (carbon- deuterium) bond is significantly stronger than the C–H bond, and therefore a cleavage of the former bond will occur several times more slowly than the corresponding cleavage of the latter bond. The common 'light' carbon atom 12C of the C–H bond can also be substituted for a heavier, stable 13C isotope, but the bond-cleavage-rate decrease will be smaller than that involving substitution of H for D because 13C is only 8% heavier than 12C. This is called the kinetic isotope effect (KIE).

An isotope is an element whose nucleus contains a specific number of neutrons, in addition to the number of protons that uniquely defines the element. It means that an atom of the same element can have different number of neutrons and a fixed number of protons. An atom with different number of neutrons are called isotope. Hydrogen has no neutrons at all. However there are some hydrogen isotopes; deuterium (hydrogen with one neutron), and tritium with two neutrons, Figure (43).

Figure (43) Isotopes of hydrogen

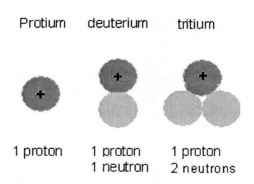

Protium   deuterium   tritium

1 proton   1 proton   1 proton
            1 neutron   2 neutrons

Isotopes are written in the form ^AB_C, where B is the symbol of the element, C is the atomic number, and A is the number of neutrons and protons combined, called the mass number. For example, ordinary Hydrogen has the formula of $^1H_1$, deuterium $^2H_1$, and tritium is $^3H_1$. Every chemical element has more than one isotope, and one isotope is more abundant in nature than any of the others. Multiple isotopes of one element can be found mixed.

Certain isotopes of elements are unstable, radiating energy or radioactive waves, and called radioisotopes. Carbon - 14 ($^{14}C_6$) is a radioisotope. Certain isotopes are more radioactive than others. There are many carbon isotopes ranging from 8C to 22C. All of them have halftimes (will be discussed soon) ranging from several seconds to microseconds, except the 14C which has 5730 years of halftime. Radioisotopes have an unstable nucleus that emits rays of types alpha, beta, or gamma (will be explained) until the element reaches stability point. Both halftimes and gamma will be discussed later in more detail.

The stability point in this case is a nonradioactive isotope of another element. For example, radium - 226 ($^{226}Ra_{88}$) will decays finally to lead - 207 ($^{207}Pb_{82}$). Precise measurement shows that some elements still contain traces of radioisotopes even with light elements such as hydrogen. Isotopes of phosphorous are used in medical therapy to kill certain cells such as cancerous cells such as leukemia (increase in white blood cells) and polycythemia (increase in red cells).

In medical diagnosis and research, isotopes of iodine are used in the diagnosis of thyroid function and in the treatment of hyperthyroidism, and in cathodic - ionic blood circulation action and reaction of substances for research. In industry, radioisotope is used in measuring the thickness of metal or plastic sheets. The testing of corrosion, oxidation, reduction, thickness of paint and other industrial process can be tested by radioisotopes.

Radioisotope releases energy (decay) by spitting out energy in the form of particles or electromagnetic waves.

2-15-1 Alpha, Beta, and Gamma Decay

Radioisotopes are continually undergoing decay.

Alpha decay is common with elements of atomic number greater than 83; from rubidium onwards. Alpha decay can be written in the following form:

$$^AB_C \longrightarrow {}^AB_C - {}^4He_2 \longrightarrow {}^{A-4}D_{C-2}$$

Note that B is the parent isotope and D is the daughter isotope and different element than element B. Helium (He) is called the alpha particle.

The new element D is now less in atomic number by 2, and consequently going back two places in the periodic table. Can we convert thallium (81) to gold (79)? Helium has two electrons in the first orbit, and they are strongly attached to the nucleus which has two positive protons. The two electrons are very difficult to leave the helium atom, yet they love to capture electrons. Therefore, alpha is not very penetrating as it captures electrons immediately from the substance it hits. However, it is very damaging because the energy released, due to the capture of more electrons, can knock atoms off the substance.

Alpha rays are used in many applications. Since alpha decay travels short distance (few centimetres or inches), it can be used to ionize small air (mainly CO_2) and allow a small current to flow such as smoke detectors. CO_2 will be ionized to CO_2^- and a current of several milliamps go through the gap of the smoke detector, sounding the alarm.

Alpha decay is used in thermoelectric generators used for space probes and artificial heart pacemakers, because alpha decay is much more easily shielded against than other forms of radiations. Plutonium -238 (the product of plutonium 244 : $^{244}Pu_{94}$ - $^{4}He_2$ = $^{238}Pu_{92}$) requires only one inch (2.54 centimetres) of lead to protect against other radiations.

Beta negative decay is the process of converting the neutron into a proton, and it follows the equation:

$$^{A}B_C \longrightarrow {}^{A}B_C - {}^{0}e_{-1}{}^{-1} \longrightarrow {}^{A}B_{C+1}$$

The tritium (H-3) can be converted into helium, Figure (44), by converting one neutron into a proton. So $^{3}H_1$ becomes $^{3}He_2$, and an unstable isotope is converted into a stable isotope of helium, or semi stable helium. For complete stability, the number of protons must equal to the number of neutrons.

Figure (44) Beta negative decay turns tritium into helium.

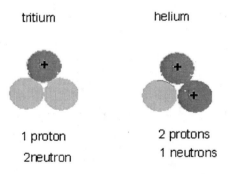

Beta negative decay is fast and furious at the beginning and slow down over time. It is more penetrating than alpha because the particles are smaller.
In general, there are three forms of beta decay:

a) Electron emission in which a neutron converts into a proton with the emission of an electron which is called a beta-minus particle. It takes the form:

$$A^0 \longrightarrow P^+ + e^-$$

Note that the total charge is the same on both sides of the formula. The electron e^- can not be existed inside the nucleus and therefore it is repelled and released to outside the nucleus.

When the electron is repelled, it could cause damage to the cell of a human body. If beta decay is not properly shielded: the electron could oxidize the human cell and cause mutation to it, and possibly a cancerous result. An example of this negative decay occurs in the iodine -131 which decays into xenon - 131 and one electron:

$$^{131}I_{53} \longrightarrow {}^{131}Xe_{54} + {}^{0}e_{-1}$$

Note that the alpha decay decreases the atomic number, whereas the beta decay increases the atomic number. In other words, Alpha is a reducing agent, and the beta is oxidizing agent. This would relate to the Warburg hypothesis for cancer growth that will be examined in chapter 3.
Above equation shows that the number of protons have been increased, and if the number of protons have increased above a certain limit, the atom will not be stable. In this case, the atom attempts to be stable again by converting some protons back to neutrons with the emission of a positively - charged electrons.

b) An electron with a positive charge is called a positron:

$$P^+ \longrightarrow n^0 + e^+$$
$$\underline{\qquad\qquad} \longrightarrow positron$$

An example of this type of decay potassium - 39 which decays into Argon -39:

$$^{39}K_{19} \longrightarrow {}^{39}Ar_{18} + {}^{0}e_{+1}$$

c) Electron capture or beta decay X-ray, in which an inner orbiting electron is attracted to unstable nucleus and combines with a proton to form a neutron:

$$e^- + p^+ \longrightarrow n^0$$

In this process, there is no radiation emitted, but the cloud surrounding the nucleus is changed, and filled by an electron from an outer shell. The filling of the vacancy is associated with the emission of an X-ray. That is called beta decay of X -ray type.

Gamma decay involves the emission of energy from an unstable nucleus in the form of electromagnetic or photon radiation. It follows the following formula:

$$^A_C B \longrightarrow {}^A_C B + {}^0_0 \gamma$$

In gamma decay, a nucleus changes from a higher energy state to a lower energy state through the emission of electromagnetic radiation or photons. The number of protons (and neutrons) in the nucleus does not change in this process, so the original and the output atoms are the same chemical element, Figure (45).

Figure (45): Gamma change in energy

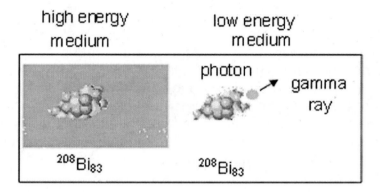

In alpha and beta decay, the atomic number of the nucleus changes, but in gamma decay the atomic number does not change. In gamma decay, the change is only in the energy state of the nucleus that will be changed to a lower state by emitting photons. The photons produced in this decay are consequently known as gamma rays and have a wavelength with an order of magnitude of about $1,000 \times 10^{-15}$ ($10^{-15} = 1$ femtometer),$= 10^{-15}$ meters. As a result, the nucleus will decay to the ground state (ground state is the condition of an atom, ion, or molecule, when all of its electrons are in their lowest possible energy level i.e., not excited. Ground state also means that all electrons fill the lowest energy orbits). By emitting one or more gamma-ray photons, Figure (46).

Figure (46): Comparison between the three isotopes in penetration

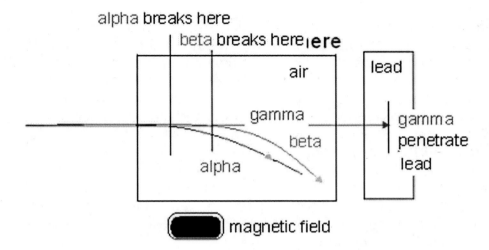

Alpha and beta decays can change the nucleus structure of an atom, i.e., the atom will be in an excited state after the decay is completed. After alpha and beta decay, gamma decay will take responsibility by releasing more energy until the atom reaches the ground state, Figure (47).

Figure (47): Staging of radioactive

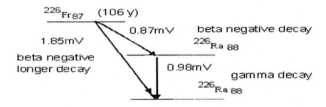

We shall show the full staging of energy and atomic number for the three decays; alpha, beta negative, beta positive, gamma in Figure (48).

Figure (48): Staging of alpha, beta negative, beta positive, and gamma decay.

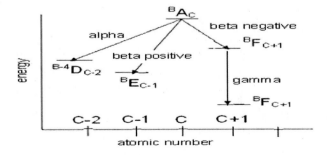

2-15-2 Half time

Radioactive atoms (unstable isotopes) are steadily disappearing, and being replaced by stable atoms. The decay from parent to daughter occurs at a certain rate. The rate of decay can be expressed as the time it takes for half of the weight of the parent to convert to its daughter. For example, the half time of beryllium 11 is 13.81 seconds. Let's start with 20 grams of ^{11}Be. After 13.81 seconds, the weight of beryllium becomes 10 grams, and the rest is converted to boron - 11. With another 1.81 seconds, we will have 4 grams of beryllium -11, and 2 grams after 13.81 seconds more, and so on, until the beryllium -11 has vanished. We are referring to Be - 11 and not Be - 9.012 which is stable. All radioactive elements disintegrate according to their specific half time, or half life. The half time of a radioisotope is the time required for half of the initial mass number of an atom to disintegrate. The following formula represents the decay time and the atomic mass of an atom:

$$Ln \frac{M}{Mo} = - kt$$

Note that M is the remaining weight, Mo is the initial weight, k is the rate of decay, and t is the time of decay. Solve the above equation, considering M/Mo equals 0.5, then ln 0.5 = - 0.693. Therefore, the half time equation can be written as:

$$t_{\frac{1}{2}} = \frac{0.693}{k}$$

Using above equations, one can predict how much of an element is left over after a certain time, and how much of the element originally existed. Therefore, ages of archaeological history of a substance, material, rocks, etc can be determined. Table (10) shows some element with their half time decay.

Table (10): Half time of some elements

Uranium	^{226}U$_{92}$	4.5×10^{9}	y
Radium	^{226}Rn$_{88}$	1602	y
Radon	^{222}Rn$_{86}$	3.82	y
Astaline	^{210}At$_{85}$	8.1	h
Francium	^{223}Fr$_{87}$	22	min
Ununbium	^{285}Uub$_{112}$	34	sec
Ununhexium	^{293}Uuh$_{116}$	5.3×10^{-2}	sec

Example: Assume there is 875 grams of radon for every 125 grams of radium as:

1000 grams of radium (initially) → 500 → 250 → 125 grams of radium

Therefore, 1000 -125 = 875 grams of radon has occurred in three half times.

The proportion of mass equals to 875 divided by 125 which equals to 7.

Thus 7 times 1602 years of half times gives a total of 11214 years have elapsed since the 1000 grams of radium converted to 875 grams of radon.

2-16 Polyatomic ions

Polyatomic ions are charged molecules in which atoms are very tightly bound together. There are three common types of polyatomic ions:

-    Families of polyatomic ions have numbers of oxygen,
-    Families of polyatomic ions have hydrogen and oxygen,
-    Families of polyatomic ions have sulfur substituted for oxygen.
-    Other families have variations of different elements bur are les common.

Let's take some polyatomic ions:

2-16-1 Carbonate or carbon trioxide CO_3^{-2}

Carbonate works as a modifier to pH in the blood. When pH is too low it absorbs hydrogen ions and causes the CO_3 to shift left, i.e. less negative. As a result, the blood will get rid of (exhale) CO_2, and decreases of H^+. When the pH is too high (alkaline), the hydrogen is reduced, and more oxygen is inhaled. Thus the pH is balanced.

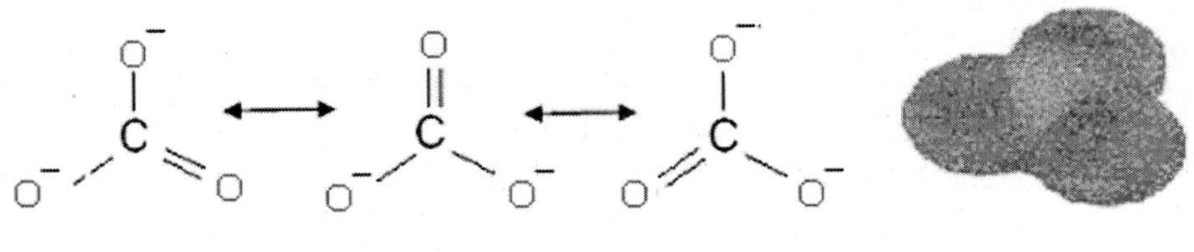

Resonance structure (will be discussed) can be used to depict the carbonate ion. The oxygen with double bond has eight electrons in the outermost orbit, but the oxygen with single bond has only seven, and therefore is negative.

2-16-2 Nitrate NO_3^-

Nitrate is toxic to humans because it oxidizes the iron in hemoglobin. The oxygen and nitrogen want to take the two and the three electrons from the two outermost orbits (14, 2) since they are very weak with respect to nitrogen and oxygen. The iron would then be double or triple cations, rendering it unable to carry oxygen as the remaining electrons are becoming stick to their last orbit. The oxygen can not pull any electrons from the iron. This condition is called methomoglbinemia.

Sometimes drinking water contains high level of nitrates. Water comes from springs close to land, and could contain high level of nitrate. Shallow rivers and creeks could have adverse effects on aquatic species.

A nitrogen atom needs only three electrons for its outermost orbit to be stable. Oxygen is more negative than nitrogen and therefore it pulls one electron from the nitrogen. Nitrogen is now positive.

2-16-3 Nitrite NO_2^-

Nitrate in food (meat and fish) delays the development of botulinal toxin that could affect the central nervous system and cause muscular paralysis. It develops cured meat flavor and color, and retards the development of rancidity and off-odors and flavors.

Nitrate can easily get rid of one oxygen atom to replace with one atom of sodium as repulsion between nitrogen and sodium is much less than that between oxygen and nitrogen. The product is called sodium nitrite, which is commonly used for curing meat products. When sodium nitrate (nitric oxide) combines with myoglobin (muscle protein), it results in the pigment responsible for the natural red color associated with cured meat. Nitrate sodium can be fatal if the dose is above the range of 23 milligrams per kilogram of the body weight. It has been reported that people normally consume more nitrates from their vegetable intake than from the cured meat products they eat. Some reports have suggested that nitrate and nitrite are related to some types of cancer diseases. However, there is no confirmed evidence in medical literatures on the carcinogenicity of nitrate or nitrite.

2-16-4 Nitronium Ion NO^+

Nitronium is a strong oxidizing agent and may cause burns to skin or eyes. It may be harmful if swallowed, inhaled or absorbed through the skin, and very destructive to mucous membranes. In vitro, it can combine with other oxidizing substances such as BO_4. The produced component can be destructive to human cells.

115pm

$$O = N^+ = O$$

180°

2-16-5 Sulfate SO_4^{-2}

Sulfur is a very oxidizing agent to magnesium. Sulfur negativity is 2.6, and magnesium negativity is 1.3. Also consider that the common valances of sulfur are -2,-1,0,+4, +6, and the common valances of Magnesium are -6,-5, -4, 0, +2). Sulphate is present in all body tissues and it combines with magnesium to play an important role in the metabolism of areas of bones and teeth formation. Sulphate loves water and any excess amount in the blood is rapidly excreted in the urinary system, dehydration can be a side effect following the ingestion of large amounts of magnesium or sodium sulphate.

Sulphate is one of the least toxic anions as it easily combines with the magnesium rather than phosphorus, or nitrogen However it can be toxic and may be lethal if the dose of magnesium sulpahate is more than 200 mg/ kg of body weight. Sulfate salts in drinking water could oxidize metal steel containers and increase corrosiveness. Sulfate can easily bond with the hydrogen in bacteria. Sulfate reducing bacteria may cause tuberculation of metal pipes, and reduce the aesthetic feature and add unpleasant odor to the water.

149 pm

2-16-6 Sulfite SO_3^{-2}

Sulfite plays an important function in the oxidation of molybdopterin as discussed in the next chapter). The lack of sulfite could lead to a disease called as sulfite oxidize deficiency. This fatal disease could cause neurological disorders, mental retardation and physical deformities due to the degradation of the brain.

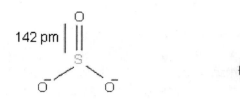

2-16-7 Hypochlorite ClO ⁻

Hypochlorite is a bond between one oxygen atom and one chlorine atom. And both atoms have negativity of 3.5 and 3.2 respectively. Since the negativity difference is not large (0.3), the attraction between them is small.Therefore the molecule is a strongest oxidizer of all chlorates, and also is the least stable. Hypochlorite can rapidly oxidize sodium (NaOCl) when reacting with sodium hydroxide solution. Sodium hypochlorite has a variety of uses and is excellent for water disinfection. It is used on a large scale for water purification, bleaching, and odor removal. When sodium hypochlorite is added to the water, it bonds with the hydrogen of the water. The acidity increased, and the pH is accordingly increased following the formula: NaOCl +H = ClOH or NaOH. The quantity of NaOH is more than NaOCl which is very corrosive.

2-16-8 Chlorite ClO₂ ⁻

Chlorite can also be easily bonded with Sodium and the product sodium chlorite, $NaClO_2$ which is used for the bleaching and stripping of textiles, pulp, and paper. It is a very strong oxidant agent, and therefore, it is used in municipal water treatment plants. Sodium chlorite should be protected from inadvertent organic substances to avoid an explosive mixture.

2-16-9   Hydrogen Carbonate (Bicarbonate) HCO_3^-

Bicarbonate is used to regulate pH in the small intestine. It is released from the pancreas in response to the call from the hormone secritin to neutralize the acidity level in the stomach. Neutralization of acids and bases can be achieved by bicarbonate. Sodium bicarbonate release CO2 when mixed with water. it is used in baking , food preparation, and sometimes to smoother a small fire.

In medical uses, it is used as an antacid to treat acid indigestion and heartburn. It can acts as an anti-fungal for dandruff caused by fungus. As a cleaning agent, it is used for cleaning and scrubbing, and for whitening teeth.

2-16-10 Phosphate PO_4^{-3}

The body needs to build and repair bones and teeth which helps nerves function, and makes muscles contract. About 85% of the phosphorus that is needed by the body is found in bones and 15% is needed by tissues and genes (DNA and RNA). Phosphate is excreted by the kidney. If the phosphate level in the body is of a high level, it means that the kidney has a problem. Creatanine and phosphate - blood test must be done to verify if a kidney has a problem. The amount of phosphate in the blood affects the level of calcium. Since the phosphorous is more negative than the calcium, they react in opposite ways. As the calcium rises, phosphate level falls, because the calcium is oxidized. Calcium and phosphate must be measured at the same time.

2-16-11 Cyanide CN⁻

Cyanide is a rapidly acting and potentially deadly chemical that exists in many forms such as hydrogen cyanide (HCN), cyanogen chloride (CNCl), sodium cyanide, or potassium cyanide (KCN). Hydrogen cyanide was used as a genocidal agent by the German in World War II, and in the Iran-Iraq War. Cyanide is contained in cigarette smoke and the combustion products of synthetic materials such as plastics and rubber.

Cyanide affects virtually all body tissues by attaching itself to metalloenzymes and rendering them inactive, or unable to carry oxygen to the brain and heart. Metalloenzymesare are those enzymes that can bind to metal such as iron, phosphorus, and chromium. Metalloenzymes are those enzymes such as iron, phosphorus and chromium that can bind to metal. Such metals can be easily combined with the anion carbon, and damage the enzyme. Note that the negativity of carbon, nitrogen, iron, phosphorus, and chromium is 2.6, 3, 1.8, 2.2, and 1.7 respectively. The lack of oxygen inhibits cellular respiration and causes cellular metabolism to shift from aerobic to anaerobic. Consequently, the organs with the highest demand of oxygen such as the brain, heart and muscles will be severely affected. Cyanide can result in ataxia (the inability to control muscle movement) and optic neuropathy that could lead to sub acute blindness.

2-16-12 Ammonium NH₄⁺

Ammonium is a less toxic form of ammonia NH3 that could be deadly depending on the pH and the temperature of drinkable water. Ammonia tends to affect the gills of fish and other water animals. In this case, where high level of ammonia exists, fish grasp at the surface of water where the oxygen level is greatest. Fish may also lie on the bottom to conserve energy, but breathe in difficulty, dying later from a lack of breathing. The most biologically important inorganic forms of nitrogen are ammonium NH_4^+, nitrite NO_2^-, and nitrate NO_3^-. All of them can be used by algae, and at high levels can be toxic to water animals. But ammonium contributes to the quality of water and to the demand of oxygen in purifying water that contain organic materials such as sewage water. Biological oxygen demand (BOD) depends on how much of the ammonium is required to balance the oxygen demand (OD) concentration in water. Biodegrading can form ammonium hydroxide (NH₄OH) which is extremely toxic to aquatic life at elevated pH levels.

2-16-13 Mercury Hg$_2$$^{+2}$

Mercury can have two cations, either Hg$_2$$^{+2}$ or Hg$_2$$^{+1}$ (or Hg $^+$). The Hg$_2$$^{+2}$ comes from two mercury atoms each having an oxidation state of one, written (1) or (I). So Hg$_2$$^{+2}$ can be written Hg (1) and called mercurous. Hg2 $^+$ can be written as Hg (II) and called mercuric. Most mercury (I) compounds are unstable, especially when exposed to light or water, or when heated. Mercurous compounds decompose on exposure to water into liquid mercury and mercuric as follows:
Hg$_2$ +2 (aq) = Hg (l) + Hg 2$^+$ (aq)

Mercurous ion can combine with chlorine to form mercurous chloride which is toxic. Most of the mercury compounds used in medicinal purposes are now illegal to manufacture or import in many countries including the U.S., Canada, Japan and Eurozone countries.

$$Hg^+ = Hg^+$$

2-16-14 Hydronium H$_3$O $^+$

Hydronium or oxonium ion is the protonation of water. Hydronium ice on oceans may be caused by ionization defects in regular ice due to the bombardment by magnetospheric particles. The implantation of protons in the ice surface, or endogenic processes in acidic oceans, or all three together, http://adsabs.harvard.edu/abs/2003AGUFM.P51B0445C.

Hydronium is a very acidic component that can burn holes in fabrics including cotton. However, it has been used extensively at pH 2.0 dilutions to treat an eye infection, mouth gargling, and teeth cleaning from any plaque buildup, http://www.altcancer.com/h3ointro.htm.

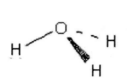 

2-17   Acids and bases

Dehydration is a common and serious problem for older adults. Normally, the level of electrolytes, acid, basis and salts do not change as a person ages; however, in the older person ICF (intercellular fluid) levels decrease by about 8% because muscle tissue changes to adipose (fat) tissue. Fat cells do not like water and therefore contain less water than muscle cells. Aged people suffer chronic diseases such as renal failure and heart failure which contribute to changes in fluid, electrolyte and acid-base balance.

The word pH stands for "power of hydrogen". It means that if a soluble compound has a lot of hydrogen, it is acidic. With less hydrogen, it is a base (alkali). How can we reduce the level of hydrogen in a compound? It is simple-just by oxidizing it (bonding it with oxygen) to become OH^- . The measurement of acids and bases is through the pH scale. The scale goes from very close to 0 through 14 as follows:

- Strong acid has a very low pH. (0 - 4)
- Strong base has a very high pH (10 - 14)
- Weak acid has a range of 3 - 6 on the pH scale. It is partially ionizes in an aqueous solution (water).
- Weak base has a range of 8 -10, and partially ionizes in water.
- Neutral has a pH of 7. It is neither acidic nor basic.

Acids are compounds that break into hydrogen H^+ ions and another compound when placed in an aqueous solution. Acids are compound that break into hydrogen H^+ ions and other compounds when placed in an aqueous solution. Bases are compounds that break into OH^- and other compounds. In other words, acids are hydrogen (H^+) donors (proton donors), and bases are oxidized hydrogen (OH^-), or hydroxide, donors. So, if an ionic compound is put in water, it will break into two ions. If the compound is acidic, more H^+ ions than OH^- will be released, and vice versa with the basic (alkaline) compound.

Fluids of the body cannot be too acidic (high level of H^+) and not too basic (high level of OH^-). The body buffers the level of both acidity and alkalinity by producing carbonic acid (H_2CO_3) or bicarbonate (HCO_3) as discussed in previous sections.

If there is an excess of acidity (H^+), then the body exhale lots of carbon dioxide:

$$H^+ + HCO_3^- \rightarrow H2CO_3 \rightarrow H_2O + CO_2$$

So, one should not drink or eat lots of acidic food, otherwise one will suffer from hypoxia that could lead to pulmonary hypertension and asthma.

If there is a shortage of H^+ then:

$$H_2CO_3 \rightarrow H^+ + HCO3^-.$$

The interpretation of pH value is:

pH 7 = 10^{-7} mol H^+ per litre or 10^{-7} mol OH^- per litre, and

pH x = 10^{-x} mol H^+ per litre or 10^{-x} mol OH^- per litre.

Some pH values are listed in Table (9).

Table (11) shows pH of some items.

Name	pH level	Molecular formula
Gastric juice	1.2 - 3.0	
Hydrochloric acid	0	HCl
Stomach acid	1	
Lemon Juice	2	
Vinegar	3	CH_3COOH
Veginal fluid	3.5 – 4	
coffee	5.0	
Milk	6 - 6.8	
Soda	4	Na2CO3, Na2O
Rain water	5	
Distilled water	7	H2O
Baking soda	9	NaHCO3
Blood	7.35 - 7.45	
Bile (Bilirubin)	7.6 - 8.6	C33H 36 N4O6
Pancreatic juice	7.1 - 8.2	
Tums (antacid) (calcium carbonate)	10	CaCo3
Ammonia	11	NH3
Mineral lime	12	Ca(OH)2

Drano (sodium hydroxide)	14	NaOH

Acids and basis can be neutralized by mixing them together to create water and salt:

HCl (acid) + NaOH (base) = H_2O (water) + NaCl (salt)

Question: Why does the oxygen, in above equation, gets rid of Na and takes instead the atom H instead. The answer is that the oxygen in NaOH has one dot of one pole is attached to hydrogen and the other dot is attached to Na, and the hydrogen of negativity of 2.1 repels the Na atom which is of negativity 0.9. Thus there is an angle of about 180 degrees between them, where as in H_2O the angle is about 104 degrees. Therefore, the repulsion between two hydrogen atoms is less than the repulsion between the hydrogen and sodium. Consequently the hydrogen atom prefers the other hydrogen atom to bind.

The same interpretation occurs with the following equation:

HBr (acid) + KOH (base) = H_2O (water) + KBr (salt)

It is important to understand the nature of acidity and alkalinity for the treatment of cancer. Cancer cells become dormant at a pH slightly above 7.4 and die at pH 8.5.
In other words, increasing alkalinity such as eating vegetation diets and drinking fresh fruit and vegetable juices can treat cancer,
http://www.alkalizeforhealth.net/cancerselftreatment.htm.
The pH scale goes from 7 to 14, and the blood, lymph and cerebral spinal fluid in the human body are designed to be slightly alkaline at a pH of 7.4. Cancerous cells can live without oxygen and love hydrogen, (i.e. acidic environments), whereas healthy tissues love alkaline environment. When oxygen enters an acidic solution it can combine with H^+ to neutralize the acidity.

In Holland, the vegetation diet promoted by Dr. Moerman has been recognized by the government as a legitimate treatment for cancer, indicating that the vegetation diet is more effective than standard cancer treatment.

Table (12): General guide for acidic and alkaline foods

Alkaline foods	Acidic foods
Beets	Olives
Broccoli	Corn

Cabbage	Blueberries
Carrot	Plums
Cauliflower	Prunes
Celery	Barely
Cucumber	Bran oat
Dandelions	Bran wheat
Eggplant	Bread
Garlic	Crackers
Green beans	Flour
Lettuce	Macaroni
Mushrooms	Oatmeal
Onions	Rice
Peas	Spaghetti
Peppers	Wheat
Pumpkins	Black beans
Radishes	Check peas
Spinach	
Sprouts	Kidney beans
Sweet potatoes	Soya beans
Tomatoes	Butter
Dandelion roots	Cheese
Apple	Ice cream
Apricot	Cashews
Avocado	Legumes
Banana	Peanuts
Berries	Pecans
Black berries	Walnuts
Cantaloupe	Bacon
Cherries	Beef
Coconut	Cod
Currants	Fish
Dates dried	Lamb
Figs dried	Lobster
Grapes	Oyster
Grapefruit	Mussels
Lemon (very alkalizing)	Pork

Honeydew melon	Salmon
Nectarine	Tuna
Orange	Turkey
Peach	Butter
Pear	Canola oil
Pineapple	Olive oil
Raisins	Sunflower oil
Strawberries	Sugar
Tangerine	Beer
Tomato	Hard liquor
Almonds	Spirits
Chesnutt	Wine
Chilly pepper	Catsup
Cinnamon	Cocoa
Curry	Coffee
Mustard	Mustard
Apple cider vinegar	Soft drink
Soured Dairy product	Aspirin
Calcium	Drugs medicinal
Cesium	Herbicides
Potassium	Pesticides
Sodium	Tobacco

2-17-1 Acidity in terms of Ka and pKa

Let's take this example:

$$HCl\ (aq) + H_2O \rightarrow H_3O^+\ (aq) + Cl^-\ (aq)$$

Acid   base   cation   anion

H_3O^+ is also called conjugate acid
Cl^- is also called conjugate base

$$Ka = [H_3O+][Cl-]/[HCl \times H_2O]$$

Note that the unit of numerators and denominators is concentration

$$pKa = -\log_{10} ka$$

Ka is called an acid dissociation constant or acidity constant or acid - ionization constant. It is a measure of the strength of an acid in a solution. If the value of Ka is larger, the stronger the acid. Stronger Ka values cover a wide range of 10^{+10} for the strongest acid such as sulfuric acid to 10^{-50} for the weakest acid such as methane. The negative sign is more appropriate to be assigned to acidity as acidity is below pH 7, and the positive sign is for alkalinity. So, pK is a more convenient scale than Ka. As an example, Ka of !0 +10 is very acid solution, but if we say pKa is -10, it is more acceptable: $pKa = -\log_{10} 10 = -10$.
Similarly, $pKa = -\log_{10} -50 = 50$ is more appropriate for alkalinity.
Generally, more negative pKa value is a stronger acid, and a more positive value of pKa is weaker acidity, or has stronger alkalinity (base). The exact values of an acid must be determined experimentally. It cannot be calculated from the negativity of each element in a molecule, because it depends on the molecular structure and its functional group. We shall discuss the following example:

$$NH_4 \quad + \quad H_2CO_3 \quad \longrightarrow \quad N^+H_4 \quad + \quad HCO_3^-$$

stronger base	stronger acid	weaker acid	weaker base

The nitrogen atom in NH3 has 5 electrons in its outermost orbit, and needs only three hydrogen atoms to be stable. If the orbit is completed to 8, then the molecular NH3 finds it difficult to release its three hydrogen atoms.

Thus it is a strong base. In the case of H2CO3 which is shown below, the oxygen has two electrons bonded to the two dot- pair; one carbon and one hydrogen and both repel each other.

The force of repel to the hydrogen electron makes the oxygen to repel the hydrogen. Therefore, the molecular H2CO3 gives hydrogen the NH3.

The pKa of the carbonic acid (H2CO3) is 6.36 and the pKa for the ammonium (NH3) is 9.24.

2-18 Acetylene series

Acetylene series is a group of aliphatic (carbon atoms joined in a string open chain) hydrocarbons, each containing at least one triple carbon bond. The group resembles acetylene and has the formula of C_nH_{2n-2}, with acetylene being the simplest formula. There are mainly four groups of acetylene that emerges from the chain. Generally, hydrocarbons are compounds that only contain H and C atoms of the formula C_nH_m, but they can be subdivided into four main groups, as shown in Figure (49).

Figure (49): groups of acetylene series

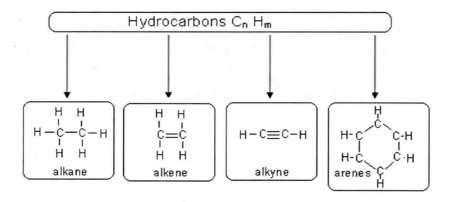

2-18-1 Alkanes

The general formula of alkane is C_nH_{2n+2}. The simplest is methane which is CH_4. Here are the first four groups of alkane.

Hydrocarbons which contain only single bonds are called alkanes. They are called saturated hydrocarbons because there is hydrogen in every possible location. This gives them a general formula of C_nH_{2n+2}, Figure (50).

Figure (50): Methyle groups of ethane and butane

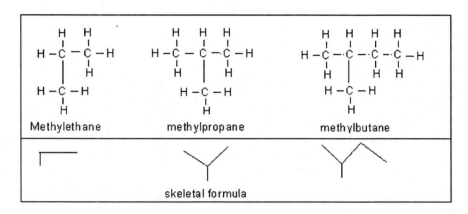

Methane can be added to ethane, propane, and butane to form methyl ethane, methyl propane, and methylbutane as follows, Figure (51).
Let's take the aliphatic heptanes, 2-chloroheptane, and 3-chloroheptane configurations, Figure (51).

Figure (51): Configuration of aliphatic heptanes

heptane

2-chloroheptane

3-chloroheptane

When carbon is double bonded with another carbon it is called alkene, Figure (52), which will be discussed in the following section.

Figure (52): Alkene configurations

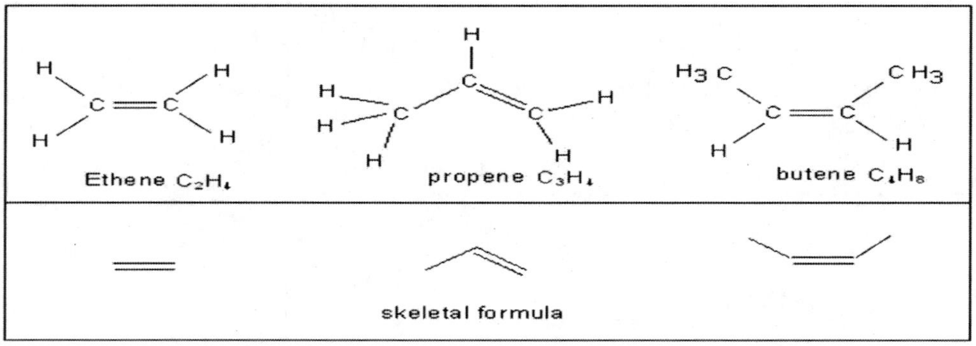

188

Ethene C_2H_4 propene C_3H_4 butene C_4H_8

skeletal formula

The suffix - ane is associated with the four groups of alkane which are gases. The prefixes of penta, hexa, hept, oct, non, and dec are used for groups 5, 6, 7, 8, 9, and 10 are for liquids. Liquids are up to $C_{17}H_{36}$. Alkanes are highly combustible clean fuels, forming heat, water, and carbon dioxide. Gasoline is a mixture of alkanes of C_5 to alkanes of C_{10}. Alkanes of C_{18} and above are solid at room temperature and are found in petroleum jelly, paraffin wax, motor oils and lubricants. Asphalt consists of a very high number of carbons.

2-18-2 Alkenes

Alkanes are one type of unsaturated hydrocarbons, because carbon atoms are with one or more double bonds, and therefore are holding fewer hydrogen atoms than they would if the bond was a single bond. Alkenes take the formula of C_nH_{2n}. Here are some alkenes.

When alkenes have more than three carbons, they are isomers. This means that there are two or more different structural formulas that can be drawn for each formula. For example, the molecule of butane has the following different structures, Figure (53).

Figure (53): Butane groups

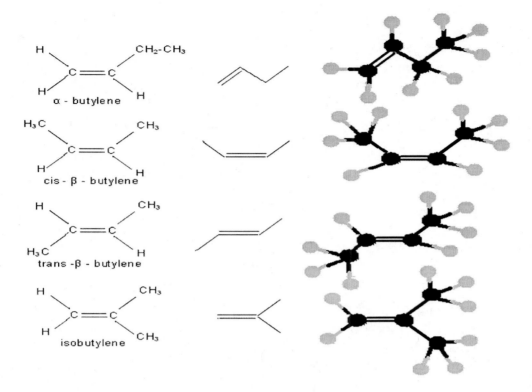

The double bonds between carbon atoms, which are relatively weak, alkenes therefore needs stronger acids to the double bonds of the carbons such as HCl, HBr, H2SO4, HI, etc. The majority of these reactions are exothermic, i.e. the output heat is higher than the combined heat of the reactants as shown in Figure (54).

Figure (54): Exothermic heat compared with reactants' heat

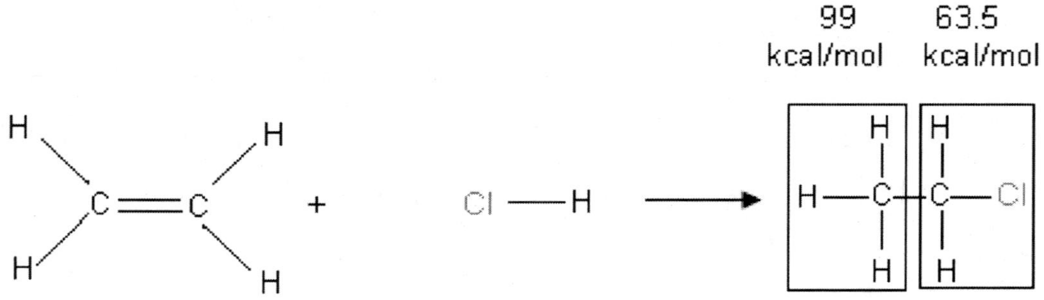

$$99 \quad 63.5$$
$$\text{kcal/mol} \quad \text{kcal/mol}$$

63 kcal/mol $\quad + \quad$ 72.5 kcal/mol $\longrightarrow$

Exothermic energy $= (99 + 63.5) - (63 + 72.5) = 27$ kcal/mol

Weak acids such as water ($pK_a = 15.7$) and acetic acid ($CH3COOH$), known also as ethanoic acid ($pK_a = 4.75$) do not normally react with alkenes. However, the addition of a strong acid such as sulfuric acid ($H2SO4$) serves to catalyze the addition of weak acids (water), to prepare alcohols ($C_nH_{2n+1}OH$) from alkenes as shown below:

$H_2SO_4 + H_2O = H_3O^+ + HSO_4$
$H_3O^+ + C_2H_4 = C_2H_6OH + H^+$

The formulas above show that sulfuric acid which is a strong acid ($pKa = -1.74$) hydrates and oxidize the ethane (ethylene) to produce alcohol.
Stronger acids such as hydrochloric acid HCl ($pka = -3$), and perchloric acid $HClO4$ ($pKa = -7$) should not be used with water, because the products will be alcohols, ether, and halide anions and other hexagonal products (benzene). Let us assume that water is used with HCl as below:

$HCl + H_2O = H_3O^+ + Cl^-$
$H_3O^+ + 2C_2H_4 + Cl^- = C_2H_6 + OC_2H_4 + Cl^- + HO^-$ $\qquad$ or
$H_3O^+ + 3C_2H_4 + Cl^- = C_6H_6 + 4H_2 + Cl^- + HO^-$

2-18-2-1 Bonding with alkenes

The double bond between the carbon atoms is two pairs of shared electrons. One of the pairs is held on the sigma bond as shown in Figure (55). The other pair is held in a molecular orbital above and below the plane of the first pair and is called the pi bond. The sigma pair is a strong bond, where as the pi bond is a weak bond and is free to move anywhere around any where around the molecule, and from one half to the other. The pi bond is relatively open to attack by other molecules and atoms, because it is exposed above and below of the molecule. As an example, let a simplest alkene is bonded to a general molecule AB, the configuration will be as in Figure (55).

Figure (55): Alkene with sigma bond and pi bond

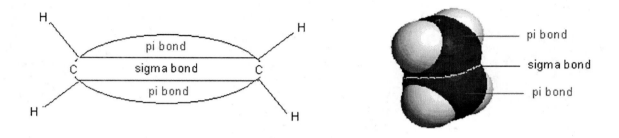

Figure (56): Addition reaction to alkene

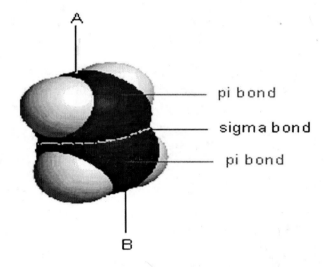

The reaction of alkene C_2H_4 is accomplished in two steps:

$$C_2H_4 + AB = AC_2H_4^+ + B^- \qquad \text{step1}$$

$$AC_2H_4^+ + B^- = ACH4B \qquad \text{step 2}$$

An energy versus progress of reaction diagram is shown in Figure (57). In step 1 the energy increases due to the disassociation of the double bond between the two carbon atoms ($\Delta E1$), and then slow down due to the production of B^- anion ($\Delta E2$). The combination between A^+ and B^- produces energy again to $\Delta E3$ and then drops in total to $\Delta E4$. The resulting energy is - ΔE.

Figure (57): Energy versus progress of reaction in alkenes

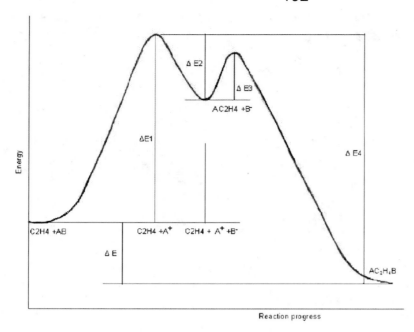

The reaction of step 1 and step 2 is demonstrated in Figure (58) below. The hydrogen atom splits the pi bond into a separate bond, combined with one carbon atom to leave the chlorine atom in an anionic charge. In the final stage of reaction the chlorine atom loses its charge and the molecule becomes stable.

Figure (58): Reaction between stable and non-stable molecule

2-18-3 Alkynes

Alkynes are hydrocarbons that have the formula of C_nH_{2n-2}. It has at least three bonds between the two carbons. Traditionally, alkynes are known as acetylenes or acetylene series. The simplest acetylene is the ethyne C_2H_2 which has the formula and shape of:

Alkynes tend to be more electropositive than alkenes, since it has two hydrogen atoms and three bonds which make the space to be attacked larger than the space of alkenes. The two hydrogen atoms are less positivity than four hydrogen atoms (hydrogen is a proton). Alkynes therefore tend to bind more tightly to a transition metals (columns 3 -12 in the periodic table and all of them end with I or 2 electrons in the outermost orbits) than alkenes. A triple bond is unstable and so alkynes are quite reactive. Since they have less hydrogen than alkenes, they are still able to act as acids. They are highly volatile, and can combust readily as propyne which is being used as a rocket fuel. Let's obserev the reaction of alkyne with a transition metal that ends with two electrons in its outermost orbit.

Figure (59) shows the reaction with a metallic base. The alkyne converts to alkene and then to alkane.

Figure (59): Alkane produced from alkyne in two steps with the use of a transition metal base

In general, alkyne that has a transitional metal reaction is resonating to either alkene or alkane, as shown in Figure (60),
http://www.ilpi.com/organomet/alkyne.html

Figure (60): Resonating alkene or alkane

2-18-4 Arenes

Arenes are aromatic hydrocarbons (compounds based on benzene rings) such as benzene and methylbenzene. Arenes are aromatic hydrocarbons, i.e. pleasant smells. They are based on benzene ring which has the simplest form of C_6H_6. The next simplest is methylbenzene (old name: toluene) which has the form $C_6H_5CH_3$. Benzene has two forms of structures; the form Kekulé and the new model form. In Kekulé form (the old one) the carbons are arranged in a hexagon, and bonds are alternating double and single between carbons. Kekulé form looks like ethene as it has double bonds between alternative carbons and one may think that benzene has reactions like ethene. Benzene is usually undergoes substitution reactions in which one of the hydrogen atoms is substituted by another atom from another substance. Benzene is a combination of 3 ethenes, and the difference between benzene and ethane is that ethane undergoes electrophilic carbon reaction and benzene undergoes substitution reaction, Figure (61).

Figure (61): Ethene with addition reaction and benzene with replacement reaction

$$CH2 = CH2 + HBr \longrightarrow CH3\overset{+}{C}H2 \longrightarrow CH3CH2Br + H$$

195

Benzene or benzol is a known carcinogen, though it is used as an additive in gasoline, is now limited, It is also used as an important solvent and precursor in pharmaceutical drugs, plastics, rubber, and dyes. It is a natural component of crude oil and smoking and may be synthesized from other compounds. Breathing benzene can cause dizziness, drowsiness and unconsciousness. Long - term benzene exposure could cause effects on the bone marrow and cause anemia and leukemia. Long term exposure to high levels of benzene in the air can cause myelogenous leukemia, which is a cancer of the blood forming organs. Benzene has been determined as a known carcinogen by the Department of Health and Human Services (DHSHS), and the International Agency for Research on Cancer (IARC),and the Environmental protection Agency (EPA).

The structure of benzene has three pi bonds, and three sigma bonds as shown in Figure (62). The sigma bonds are shown in white color lines in the structure of benzene. The picture can be redrawn in different shapes; with a doughnut above and below the horizontal ring to represent the delocalized electrons.

Figure (62): Three pi bonds and three sigma bonds of benzene

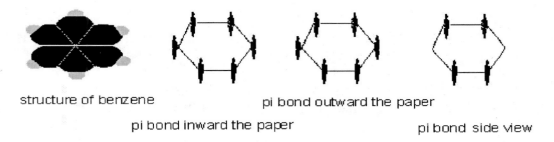

structure of benzene   pi bond outward the paper

pi bond inward the paper   pi bond side view

Normally the length of single and double carbon bonds is of different length.

C - C bond length   0.154 nm
C = C bond length   0.134 nm

If we consider the Kekulé structure, it means that the benzene (hexagon) is not regular with the alternating longer and shorter sides. Actually, in benzene all the bonds are of same length, and equal to 0.139 nm. Not only the length is different, but the Kekulé structure also gives the benzene more energy than the real benzene, which is not correct. Let's see the difference in the thermo chemistry between the Kekulé structure and real benzene by adding hydrogen (hydrogenation) to cyclohexane which is a ring of six carbon atoms containing just one double bond as in Figure (63) below.

Figure (63): Energy levels with single, double, and triple (benzene) two bonded carbons

$$\text{(cyclohexene)} + H_2 = \text{(cyclohexane)} \quad -120 \text{ kJ/mol}$$

$$\text{(cyclohexadiene)} + 2H_2 = \text{(cyclohexane)} \quad -240 \text{ (232 actual) kJ/mol}$$

$$\text{(benzene)} + 3H_2 \quad \text{(cyclohexane)} \quad -208 \text{ kJ/mol}$$

The enthalpy change during the reaction from single double bond cyclohexane is -120 kJ/mol, i.e. 120 kJ/mol of heat is evolved.

Note that in Figure (64) above, the enthalpy is conventionally written in minus, although the temperature is released from the reactant (the double bond cyclohexane).

In the double two bonded carbon (middle one above), the temperature is expected to be 240 kJ/mol, and it is actually 233 kJ/mol.

In the third one (benzene), the temperature is not 360 as one would expect, it is 208 kJ/mol, http://www.chemguide.co.uk/basicorg/bonding/benzene1.html#top

It is important thing to notice is that real benzene is much lower than the Kekulé structure would expects. This means that real benzene is 150 kJ/mole more stable than the Kekulé structure, and this decrease in energy (increase in the stability of benzene) is known as the delocalization energy or resonance energy, see Figure (64) below.

Figure (64): The lower down energy is the more energetically stable

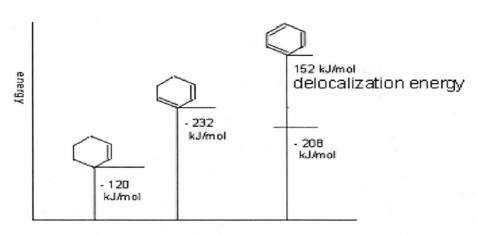

152 kJ/mol
delocalization energy

- 232 kJ/mol

- 208 kJ/mol

- 120 kJ/mol

2-19 IUPAC nomenclature

The International Union of Pure and Applied Chemistry (IUPAC) have developed a set of rules for giving a unique name for each organic compound. For example, alkanes come in all shapes and sizes. The simplest alkane has all of its carbons chained together in a series (row). Here is a row of five carbons:

$CH_3 - CH_2 - CH_2 - CH_2 - CH_3$

Pentane has two parts to its name- pent means five, and ane means the compound alkane.

Table (13) shows the appropriate base part for other number of carbons.

Table (13): Number of carbons and base

198

Table (11) Number of carbons and base

Number of carbons	Base	Number of carbons	Base
1	Meth	11	Undec
2	Eth	12	Dodec
3	Prop	13	Tridec
4	But	14	Tetradec
5	Pent	15	Pentadec
6	Hex	16	Hexadec
7	Hept	17	Heptadec
8	Oct	18	Octadec
9	Non	19	Nonadec
10	Dec	20	Icos

Now you know the twenty names of alkane bases, then you can name other number of carbons, Figure (65).

Figure (65): Anes of carbon

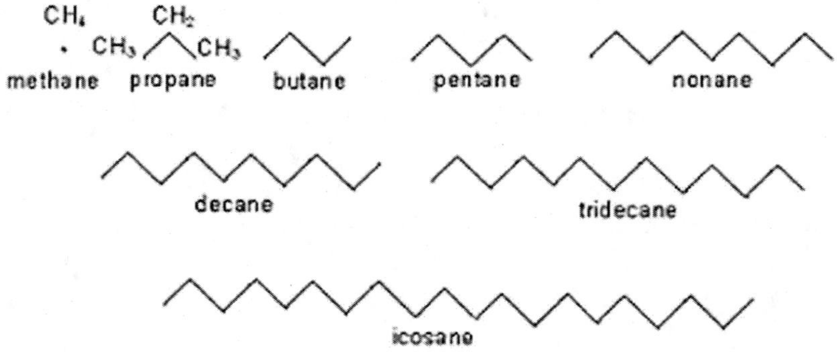

One can see that alkane can still have the same number of carbons even the chain is not on the same horizontal line as shown with tridecane above. Horizontal lines can also be represented in rings which do not have CH3 anywhere if the molecule is stable as shown below, Figure (66).

Figure (66): Different functional groups of alkanes

199

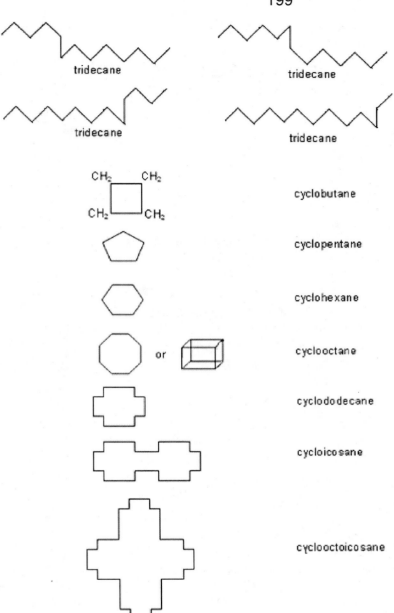

2-19-1 Numbering of IUPAC (International Union of Pure and Applied Chemistry)

Numbering can start from left or right depending on the location of other carbon atoms attached to the chain as indicated in Figure (67).

Figure (67): Correct and incorrect numbering of molecules

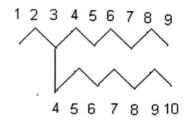

9 8 7 6 5 4 3 2 1

8 9 10 11 12 13 14

This is correctly numbered
because it is the longest chain

1 2 3 4 5 6 7 8 9

4 5 6 7 8 9 10

This is not correctly numbered
because it is not the longest chain

Figure (68) is again for nonane because it has nine carbons, but has an attachment of CH_3. The base for one carbon is meth and the attachment is defined by IUPAC as yl, so the connected branch to the chain is called methyl, so we have a methylnonane. We can show the three types of the connection yl as in Figure (68).

Figure (68): Three types of connection to the chain nonane

1 2 3 4 5 6 7 8 9

3-ethyl-4-methylnonane
correct

1
2
3
4
5
6 7 8 9

2-methyl-3-ethylpentane
incorrect

2
1
3
4
5
6 7 8 9
6
7

3-ethyl-methylethyl nonane   correct
3-methyle-5-ethyleheptane   incorrect

Groups can also be attached to rings and rings and rings can be attached to other rings, See the following groups:

Let's consider the following two structures of the same number of carbons and hydrogens. The one on the left has the two carbons are attached to the middle carbon which is attached to the cyclohexane. The one on the right has three carbons are attached in a series, and the end of the chain (CH_2) is attached to the cyclohexane. IUPAC names the nodal connection as iso and the series as a straight name.

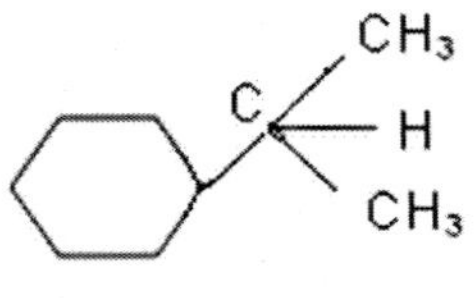

isopropylcyclohexane

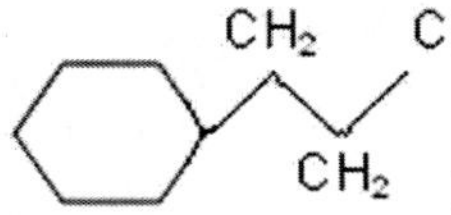

propylcyclohexane

The nomenclature above is not difficult if we use a step by step method for numbers and names. We shall treat some other names which are called systematic names which gives numbers for locations of attached molecules. One has to remember that nodal connection is not a stable molecule as is the straight chain-connection. The molecule with more nodal connections is more volatile and has a lower boiling point than molecules with less nodal connection, providing that both have the same number of carbons and hydrogens. Figure (69) has more complicated names.

Figure (69): Common names and systematic names

	common names	systematic names
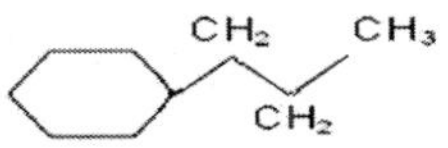	n-propylcyclohexane n stands for normal	propylcyclohexane
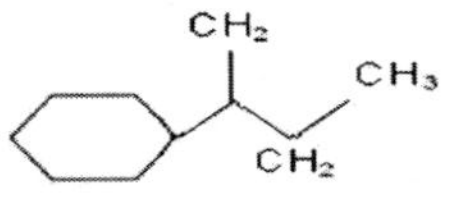	sec-propylcyclohexane sec stands for two carbons to the carbon attached to the cyclohexane	(1-methylethyl)cyclohexane
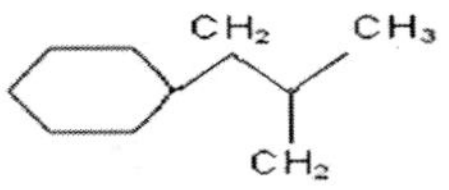	isopropylcyclohexane iso not in Italic	(2-methylethyle)cyclohexane
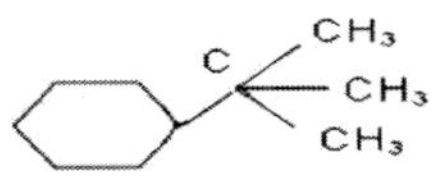	tert-butylcyclohexane tert from tertiary means 3	(1,1-dimethylethyl)hexane di means 2

If the side chain is more than 4 carbons, it is commonly named systematically. Organic compounds can have more than one group attached to the longest chain. Table (14) has names for simple groups prefix and complex group prefix.

Table (14): Prefixes of simple and complex groups

Number of groups	Simple group prefix	Complex group prefix
2	di	bis
3	tri	tris
4	tetra	tetrakis
5	penta	pentakis
6	hexa	hexakis
7	hepta	heptakis

Let's look at the examples as shown in Figure (70).

Figure (70): More than two simple groups on the main chain

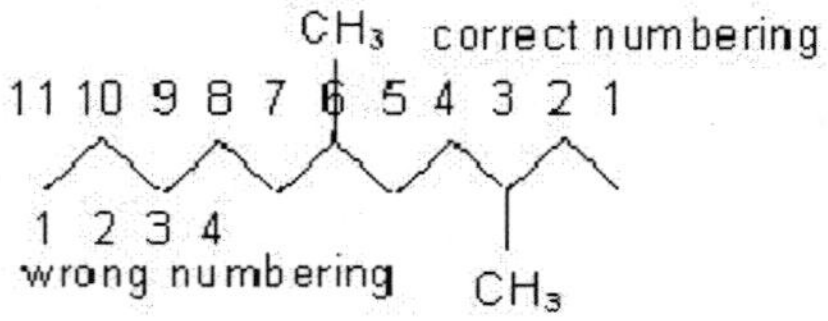

3,6-dimethylundecane

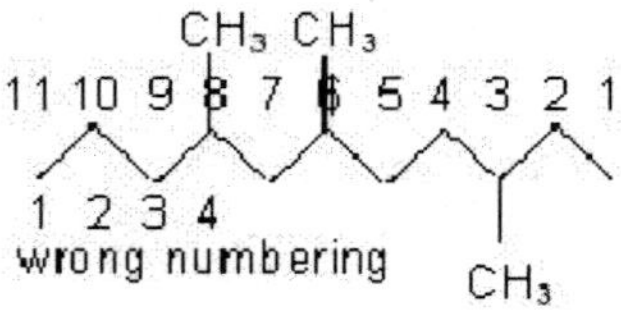

3,6,8-dimethylundecane

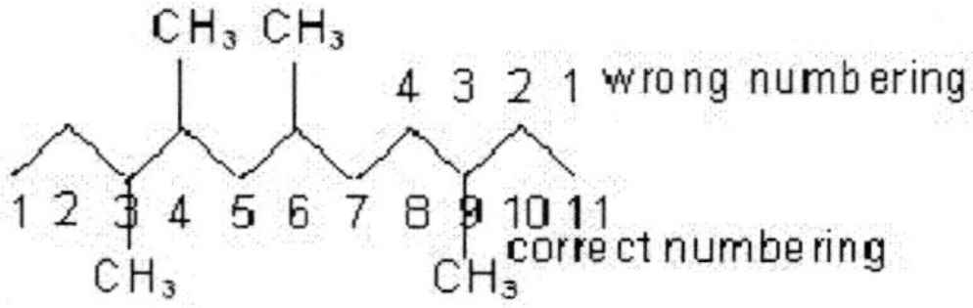

3,4,6-dimethylundecane
not 3,6,8-dimethylundecane

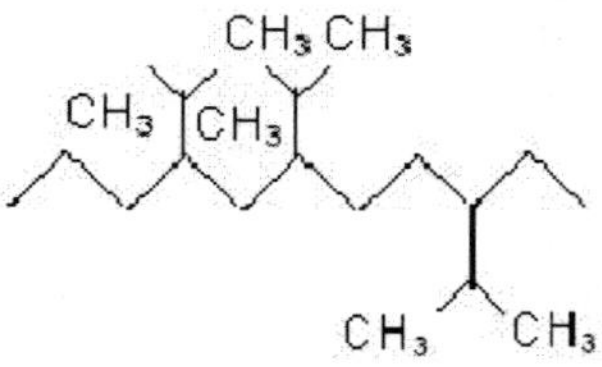

3,6,8-triisopropylundecane
or 4,6,9- triisopropylundecane
or 3,6,8-bis(1-methylethyl)undecane)
or 4,6,9-bis(1-methylethyl)undecane)

In Figure (70), the top configuration has the numbering from right to left, and the methyl hits carbon at number 3. Numbering can not be from left to right as the methyl hits the main chain at carbon number 6. In configuration number two from the top, numbering should also be from left to right for the same reason as explained above. In the third configuration, numbering is correct from right to left as the methyl hits the carbons at 3,4,and 6, not at 3,6, and 8. The fourth configuration can be from either sides because the group propyl or (methylethyl) hits the main chain at 3,6,8 or 4,6,9 or (3,6,8 or 4,6,9) respectively, the difference between the hitting points are the same.

Consider the connections of rings in Figure (69). There is no start or end where we can number the connections. We number starting at the carbon that one of the groups is connected to. Then numbering is done in the direction of the closest group. In Figure (71), the cyclohexane is connected to three groups of methane. So, the nomenclature as per the IUPAC is 1,2,4-trimethylcyclohexane. Therefore, the green numbering is wrong. Note that three connections are supposed to be horizontal to the ring. If the connections are inward to the paper plane, then the name of the component is preceded with cis-, and if the connection is outward, it is preceded with trans-. Both cis and trans are in Latin.

Figure (71): Numbering a prefix of rings with connected groups

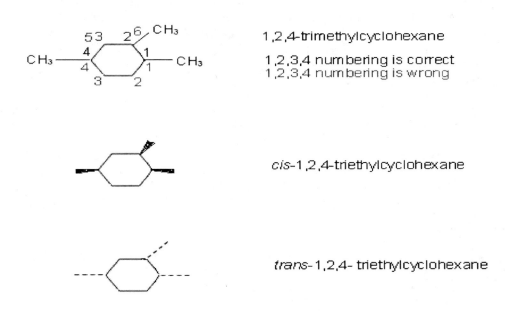

1,2,4-trimethylcyclohexane

1,2,3,4 numbering is correct
1,2,3,4 numbering is wrong

cis-1,2,4-triethylcyclohexane

trans-1,2,4- triethylcyclohexane

Now, if you understand the previous demonstration of the IUPAC nomenclature, you can name hundreds of groups. However, there are still many complicated molecules needed to be explained so that one can understand the whole subject., see Figure(72).

Figure (72): More complicated chains and rings

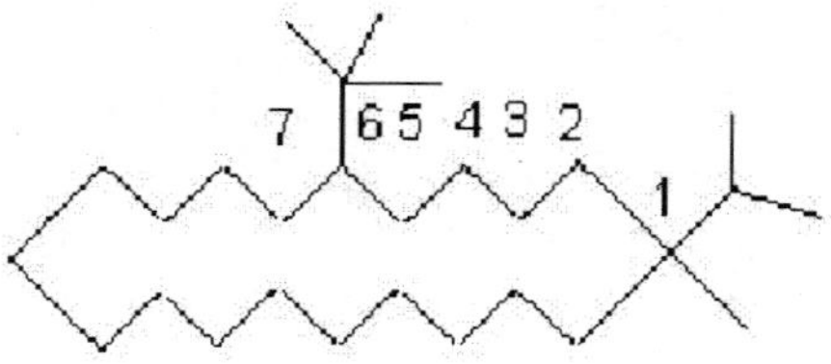

1-methyl-1-isopopyl-6-isobutylcycloicosane or
1-methyl-(1-methylethyle)-6-(1-ethyl-1-isoethyl)cycloicosane or
1-methyl-(1-methylethyle)-6-(1-ethyl-1methyl-1-methyl)cycloicosane.

2-19-2 Functional Groups

An atom or group of atoms, that replaces hydrogen in an organic compound and that defines the structure of a family of compounds and determines the properties of the family, http://www.answers.com/topic/functional-group

Alkanes, alkenes, alkynes, and arenes are functional groups. In addition, there are about 30 or so functional groups that determine the properties and reaction chemistry of molecules. It is essential for professional chemists, to be able to name organic molecules, predict solubility, chemical reactivity, and the spectra of drug effectiveness to recognize the functional groups. For example, drug morphine may have several functional groups as shown in Figure (73), http://www.chemistry-drills.com/functional-groups.php?=simple

Figure (73): Morphine drug with several functional groups

methylnitrene ($3°$amine)

phenol

allylic alcohol

ether

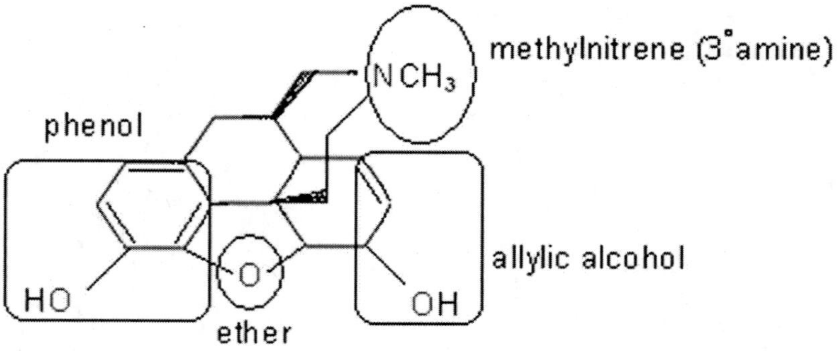

molecule morphene

205

Table (15) lists the functional groups of interest to understand the aim of this book.

Table (15): Functional groups

Class of compound	Functional group	IUPAC name	Formula
halide	-F fluoro -Cl chloro -Br bromo -I iodo	R-X R represents alkyl, X represents halogen 2-chloropropane	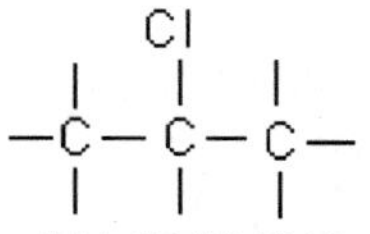 $CH_3CHClCH_3$
primary alcohol	-OH	$R=CH_3CH_2$ 1-propanol	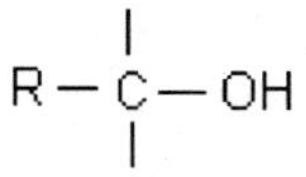 RCH_2OH
Secondary alcohol	-OH	$R_2= 2(CH_3CH_2)$ 1-diethylisopentane	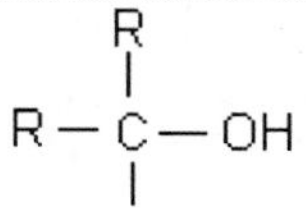 R_2CHOH
Tertiary alcohol	-OH	$R_3=3(CH3CH2)$ 1-triethyleisoheptane	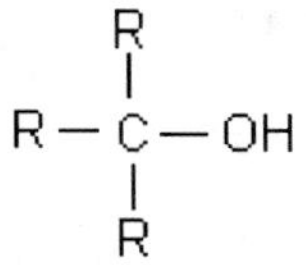 R_3COH
ether	-O-	ethylmethyether or methylethylether	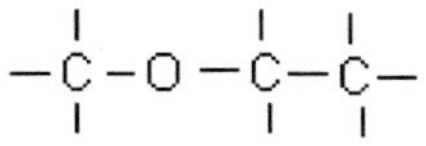 $CH_3OCH_2CH_3$
aldehyde	$-\overset{O}{\overset{\|\|}{C}}-H$	propanal	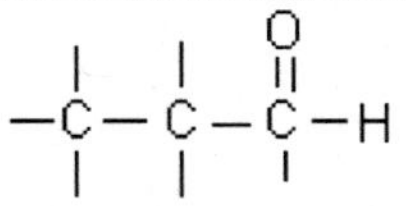 $CH3CH2CHO$
ketone	$-\overset{O}{\overset{\|\|}{C}}-$	2-pentanone	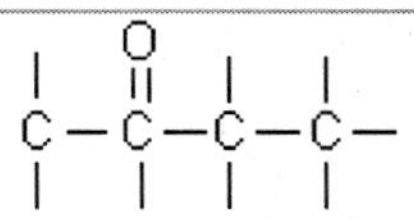 $CH_3COCH_2CH_2CH_3$
organic acid	$-\overset{O}{\overset{\|\|}{C}}-OH$	propanoic acid or ethanecarboxylic acid	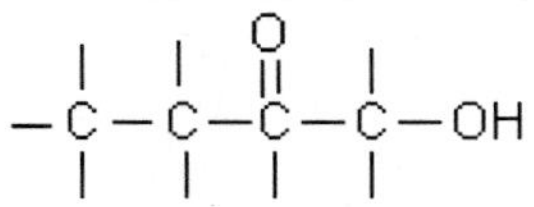 $CH_3CH_2\,COOH$

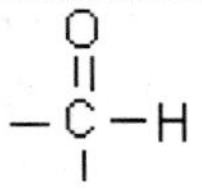

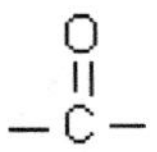

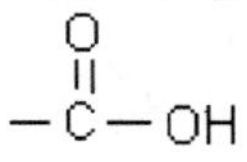

Name	Functional Group	Naming	Example
ester	O‖ —C—O—	double bond oxygen methylpropanoate single bond oxygen	O‖ —C—C—C—O—C— $CH_3CH_2\ COOCH_3$
amine	—N—	1-propanamine	—C—C—C—N $CH_3CH2CH_2NH_2$
amide	O‖ —C—NH	propanamide	O‖ —C—C—C—N $CH_3CH_2CONH_2$
acid chloride	O‖ —C—Cl—	propanoyl chloride oyl replace e of propane when connected to oxygen O	O‖ —C—C—C—C—Cl $CH_3CH_2CH2OCl$
acid anhydride	O‖ O‖ —C—O—C—	ethanoic anhydride	O‖ O‖ —C—C—O—C—C— $CH_3COOCOCH_3$
nitrite	—C≡N cyano group	Ethanenitrile or acetonitrile	—C—C≡N CH_3CN
Amino acid alanine	O=C—OH —C—NH₂	2-aminopropanoic acid	O=C—OH —C—NH₂ —C— $CH_3CH(NH_2)COOH$
Amino acid isoleucine		2-amino-3-methylpentanoic acid	O=C—OH —C—NH₂ —C—CH₃ —C— CH₃ $CH(NH_2)CH(CH_3)CH_2CH_3COOH$

tyrosine		2-amino-3-(4-hydroxyphenyl)-propanoic acid	 $OHC_6H_4C_2CHH_2NCOOH$
trans alkene	$R_2C_2H_2$	alkenyle lithium	RC_2H_2Li
trans-alkene		trans-3-methylhex-3-ene	C_6H_{11}
cis alkene	$R_2C_2H_2$	cis-3-methylhex-3-ene	C_6H_{14}

2-19-3 IUPAC Names of Medicinal Drugs

2-19-3-1 Amoxillin (antibiotic)

7-

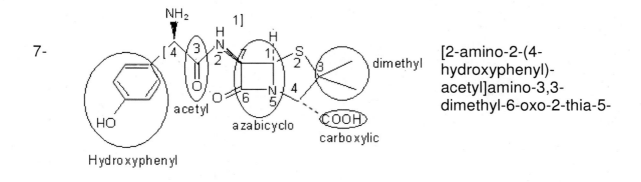

[2-amino-2-(4-hydroxyphenyl)-acetyl]amino-3,3-dimethyl-6-oxo-2-thia-5-

azabicyclo[3,2,0]heptane-4-carboxylic acid
http://en.wikipedia.org/wiki/Amoxicillin

2-19-3-2 Valium (anxiety),

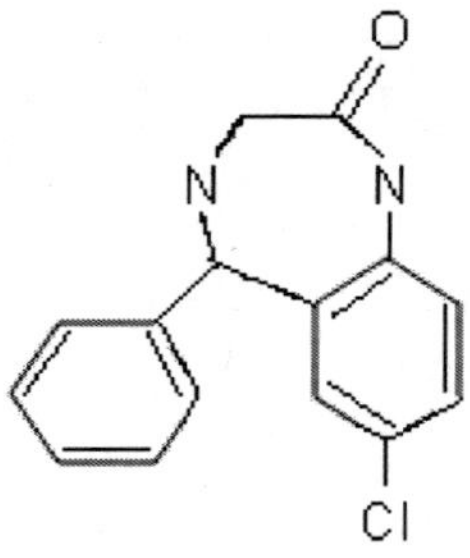

7-chloro-1-methyl-5-phenyl-1,3-dihydro-2H-1,4-benzodiazepin-2-one
http://en.wikipedia.org/wiki/Valium

2-19-3-3 Benadryl (Antihistamine)

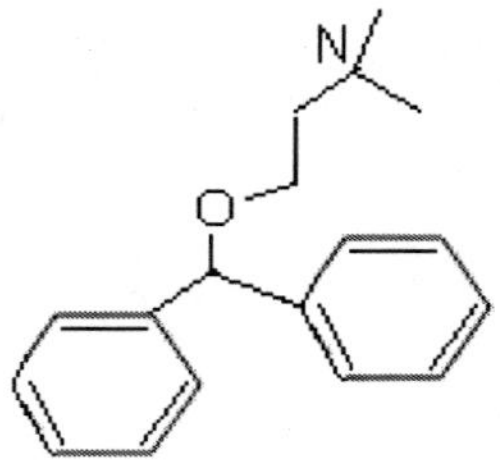

2-benzhydryloxy-N,N-dimethyl-ethanamine
http://www.chemindustry.com/chemicals/543975.html

2-19-3-4 Cialis (Dysfunction)

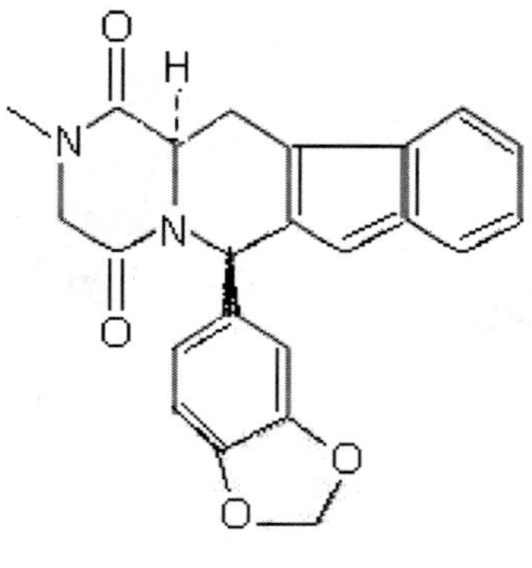

209

(6R-trans)-6-(1,3-benzodioxol-5-yl)-2,3,6,7,12,12a-hexahydro- 2-methyl-pyrazino[1',2':1,6]pyrido[3,4-b]indole-1,4-dione
http://www.medical-look.com/reviews/Cialis.html

2-19-3-5 Heparin (anticoagulant, anti-High blood pressure)

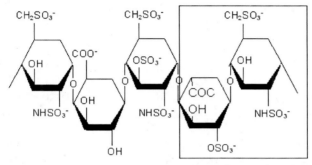

2-deoxy-2-sulfamido-α-D-glucopyranosyl-6-O-sulfate

2-19-3-6 Ibuprofen sodium (anti-inflammatory)

Sodium 2-(4-isobutylphenyl) propionate
http://pubchem.ncbi.nlm.nih.gov/summary/summary.cgi?sid=739221

2-19-3-7 Aspirin (aches and pains)

2-(acetyloxy)benzoic acid
http://en.wikipedia.org/wiki/Aspirin

2-19-3-8 Vitamin C (Scurvy)

2-(1,2-dihydroxyethyl)-4,5-dihydroxy-furan-3-one
http://www.chemblink.com/products/6272-26-0.htm

2-19-3-9 Tylenol (fever and headaches)

N-(4-hydroxyphenyl)acetamide
http://www.chemindustry.com/chemicals/512597.html

2-19-3-10 Warfarin (Rodenticide and Anticoagulant)

2-hydroxy-3-(3-oxo-1-phenylbutyl)chromen-4-one
http://www.drugbank.ca/drugs/DB00682

2-20 Atomic orbitals

Electrons usually inhabit regions of space known as orbitals. Orbitals have quantum of size, shape and orientation. The Heisenberg Uncertainty Principle pointed out that locations and angular momentum can not be predicted, and it is impossible to know the futuristic movement of an electron orbiting the nucleus. The dots in the diagram below show the location of the electron of hydrogen atom. The electron moves from one dot to another in a 3D spherical space, surrounding the nucleus, without knowing the path that it will follow from one dot to another.

90% of the time, the electron will have a defined region of space close to the close to the nucleus where the electron exists. At each dot, the electron has electron has a specific quantum of speed, location and angular displacement. The space shown in the left diagram is for the

The hydrogen atom has only one electron. But what about the helium atom which has two electrons. Helium atom has a similar spherical orbit but filled with two electrons. Lithium atom has three electrons; two in the first spherical orbit, and on in a second spherical orbit as shown below:

The first orbit of the hydrogen atom is called the 1s orbital. The "1" means that the electron is in the energy level closest to the nucleus, and "s" is the shape of the orbital. Now "2s" as for the helium and lithium atoms means that there is a second orbit similar to the first, but in different energy level. The closer the nucleus the electrons get, the lower their energy. This is clear as energy is proportional to the radius of the orbit from the nucleus (centre). Electrons can revolve in other outsider orbits called 2s, 3s, 4s, etc of similar orbits. Each s orbit has levels of energy. For example, the 1s orbital has the first energy level. Similarly, the 2s is in the second energy level, and so on. At the second energy level of 1s, there is a second type of orbital called 2p as shown in the diagram on the right below:

The p orbital is similar to 2 identical balloons tied together at the nucleus, as shown in the cross section on the left. Now we will put together the p orbital onto the s orbital as shown on the right diagram.

There are three orientations of p orbitals; one on the X axis, the second on the Y axis, and the third on the Z axis.

The three orbitals of P together with 1s orbital are shown below:

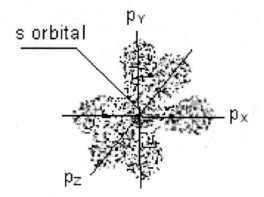

The three p orbitals start at one energy level that is mutually located at right angles to each other. There are similar orbits at subsequent levels of energy; $3p_X$, $3p_Y$, $3p_Z$, $4p_X$, $4p_Y$, and so on.

In general, there are three quantum orbital numbers that describe an orbital. They are called n, l, and m. Orbital n is the principal quantum number that determines the size of the atom. For example, the principal quantum number (n) 1 has the s orbital, number 2 has orbitals s and d, number 3 has orbitals s, p, and d, number 4 has orbitals s, p, d, and f, and so on. The second orbital number l determines the shape of the orbital. As we said before, s has the shape of a ball on the three axes; X, Y, and Z, and the p orbital has the shape of two balloons tied together on X, Y, and Z axes as shown above. The d has 5 orientations (p has 3 orientations), and f has 7 orientations. The third orbital number m_l determines the orientation, Table (16).

Table (16): Relationships between the three quantum numbers.

n	l (0,n-1)	0	1	2	3	orientation
1	s	0 (s)				s =1
2	s p	0 (s)	-1 0 1 (p)			p = 3
3	s p d	0 (s)	-1 0 1 (p)	-2 -1 0 2 1 (d)		d = 5
4	s p d f	0 (s)	-1 0 1 (p)	-2 -1 0 2 1 (d)	-3 -2 -1 0 1 2 3 (f)	f = 7

Figure (74) represents the atom with its orbitals, assuming the atom of 4 quantum sizes.

Figure (74): An atom of 4 quantum sizes

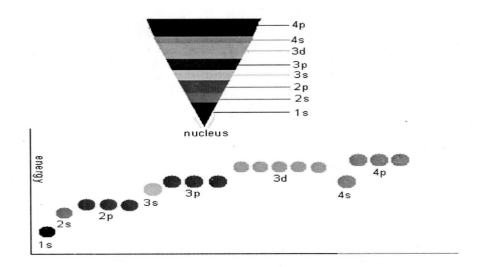

Figure (75) is not to scale but it indicates the energies of the orbitals up to the 4p orbitals. One should notice that the s orbitals have a slightly lower energy than the p orbitals of the same energy level. This means that the s orbitals always fill with electrons before the corresponding p orbitals fill their electrons. However, the three d orbitals are slightly higher than the level of 4s. Thus, the 4s orbital will fill before the 3d orbitals, and then will be followed by the 4p. Similarly the 4f orbitals will fill after the 5s. Notice that each orbit has only two electrons.

Examples

Hydrogen $1s$
Neon $1s^2 2s^2 2p_x^2 2p_y^2 2p_z^2$ or $1s^2 2s^2 2p^6$. We shall neglect orientation
Scandium $1s^2 2s^2 2p^6 3s^2 3p^6 3d^1 4s^2$
Vanadium $1s^2 2s^2 2p^6 3s^2 3p^6 3d^3 4s^2$
Copper $1s^2 2s^2 2p^6 3s^2 3p^6 3d^{10} 4s^1$
Krypton $1s^2 2s^2 2p^6 3s^2 3p^6 4s^2 3d^{10} 4p^6$
Radon $1s^2 2s^2 2p^6 3s^2 3p^6 4s^2 3d^{10} 4p^6 5s^2 4d^{10} 5p^6 6s^2 4f^{14} 5d^{10} 6p^6$

2-21 Solubility

Solubility will be affected by the polarity of both the solute and the solvent. Generally, polar solute molecules will dissolve in polar solvents and non-polar solute molecules will dissolve in non-polar solvents. For example, water is polar because of the unequal sharing of electrons. Methane is non-polar because the carbon shares the hydrogen atoms uniformly, and phospholipids are surfactant, i.e. hydrophilic and lipophilic, Figure (75).

Figure (75): Examples of polar and non polar molecules

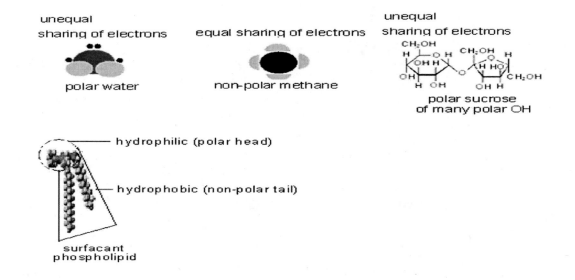

The unequal sharing of electrons within a molecule leads to the formation of two different poles (dipole); one negative and one positive.

Atoms on the right four groups of the periodic table are generally electronegative and exert a greater pull on electrons of the left groups, For example, oxygen exert greater pull on hydrogen than carbon, and fluorine exerts greater pull on hydrogen more that oxygen, and so on. Figure (76) shows that the hydrogen is closer to fluorine than oxygen, because the latter has less negativity than fluorine.

Figure (76): Hydrogen is closer to fluorine than oxygen.

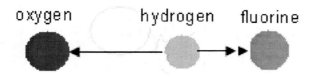

fluorine exerts greater pull on hydrogen than oxygen

In the polar solvent molecule the positive ends will attract the negative ends of solute molecule. Figure (77) shows water (polar) and boron difluoride Bf_2 (polar) and their attraction.

Figure (77): Attraction between positive and negative ions

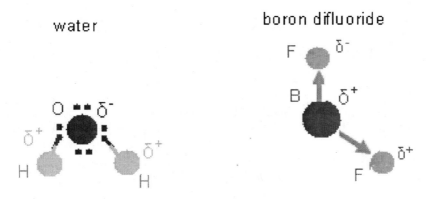

fluoride attracts hydrogen and oxygen attracts boron

2-21-1 Solubility rules

- All common salts of Group 1 elements and ammonium (NH_41^+) are soluble.
- All sulfates are soluble except barium, strontium, lead (II), calcium, silver, and mercury (I)
- All common nitrates (NO_31^-) and acetates are soluble.
- Most chlorides, bromides, and iodides are soluble except silver, lead (II), and mercury (I)
- Except for those in Rule 1, carbonates, hydroxides, oxides, and phosphates are insoluble.

Some of soluble and insoluble compounds are shown in Table (14), http://www.sciencebyjones.com/solubilty_rules.htm

Table (18): Some compounds of solubility and insolubility. Compounds that are slightly soluble are abbreviated as SS.

	acetate	bromide	carbonate	chloride	chromate	hydroxide	iodide	nitrate	phosphate	sulfate	sulfide
aluminum	ss	soluble	not exist	soluble	not exist	nearly insol	soluble	soluble	nearly insol	soluble	decomp
ammonium	soluble	soluble	soluble	soluble	soluble	soluble	soluble	soluble	soluble	soluble	Soluble
barium	soluble	soluble	nearly insoluble	soluble	nearly insoluble	soluble	soluble	soluble	nearly insoluble	nearly insoluble	decomp
calcium	soluble	soluble	nearly insoluble	soluble	soluble	ss	soluble	soluble	nearly insoluble	ss	decomp
copper (II)	soluble	soluble	nearly insoluble	soluble	nearly insoluble	nearly insoluble	decomp	soluble	nearly insoluble	soluble	nearly insoluble
iron (II)	soluble	soluble	nearly insoluble	soluble	not exist	nearly insoluble	soluble	soluble	nearly insoluble	soluble	nearly insolubl

											e
iron (III)	soluble	soluble	not exist	soluble	nearly insoluble	nearly insoluble	not exist	soluble	nearly insoluble	ss	decomp
lead (II)	souble	ss	nearly insoluble	ss	nearly insoluble	nearly insoluble	ss	soluble	nearly insoluble	nearly insoluble	nearly insoluble
magnesium	soluble	soluble	nearly insoluble	soluble	soluble	nearly insoluble	soluble	soluble	nearly insoluble	soluble	decomp
mercury (I)	ss	nearly insoluble	nearly insoluble	nearly insoluble	ss	not exist	nearly insoluble	soluble	nearly insoluble	ss	nearly insoluble
mercury(II)	soluble	ss	nearly insoluble	soluble	ss	nearly insoluble	nearly insoluble	soluble	nearly insoluble	decomp	nearly insoluble
potassium	soluble	soluble	soluble	soluble	soluble	soluble	soluble	soluble	soluble	soluble	Soluble
silver	ss	nearly insoluble	nearly insoluble	nearly insoluble	ss	not exist	nearly insoluble	soluble	nearly insoluble	ss	nearly insoluble
sodium	soluble	soluble	soluble	soluble	soluble	soluble	soluble	soluble	soluble	soluble	Soluble
zinc	soluble	soluble	nearly insoluble	soluble	soluble	nearly insoluble	soluble	soluble	nearly insoluble	soluble	nearly insoluble

CHAPTER   3

AGING

Introduction

There is no doubt that our lives can be negatively affected by chemical and environmental pollutants, stress, and even the diet. In North America, diet is rich with acidosis which may be the leading cause of many common degenerative diseases. Maintaining a normal pH balance is the first line defense against aging and diseases. For example, if a pH imbalance continues long enough, there can be a depletion of bone mass which can lead to an osteoporosis disease. Here are few diseases due to acidic body:

- Excessive stress
- Mental confusion
- Weak kidney
- Weight gain

Other side effects include malaise, headache, constipation, and cold

If the pH of the blood drops, ketone bodies will be increased resulting in ketonacidosis. Ketone bodies are produced as by-product when fatty acids are broken for energy in the liver and kidney. If ketone bodies in the blood is high, then fat will not be broken down, and it will be stored and accumulated in the liver.

3-1   Ketoacidosis

Ketoacidosis involves the buildup of toxic substances called ketones that make the blood too acidic. High ketone levels can be readily managed, but if they aren't detected and treated in time, a person can eventually slip into a coma and die. The chemical formula of a ketone molecule is shown in Figure (1).
Diabetic ketoacidosis (DKA) may occur in people who have type 1 (insulin-dependant) diabetes. A high percentage (80%) of patients with type one diabetes having ketoacidosis and low blood glucose are admitted to hospital for emergency treatment. In ketoacidosis, the body fails to adequately regulate ketone production causing such a severe accumulation of keton acids, which makes the pH level of the blood decrease. If the pH is below 3, it could be fatal. Ketone molecules like acetoacetate, beta-hydroybutyrate and acetone are released into the blood from the liver when hepatic lipid metabolism has changed to ketogenesis which is increased due to insulin deficiency. Ketoacidosis (ketogenesis) is a high anion acidosis, which is excessive hydrogen atoms. That is why the pH is becoming low, as per the following flow chart:

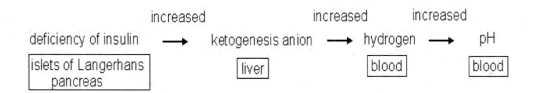

Figure (1): Some types of Ketone formula

2-butanone = ketone

3-pentanone = ketone

3-methyle-2-butanone = ketone

methyle propyle ketone

The amount of ketones in the urine should be checked if a person with diabetes has:

- Fasting blood glucose levels above 14 mmol/L (252 mg/dL) for 24 hours
- An infection or illness

Testing ketone level can be done by test-strips which indicate the level of β-hydroxybutyrate. This is done using dipsticks coated in nitroprusside or similar reagents. Nitroprusside changes from pink to purple in the presence of acetoacetate (ketone) the conjugate base of acetoacetic acid. The popular dipstick used to detect ketone bodies in urine is "Ketostix" by Bayer. It only detects acetoacetate, not beta-Hydroxybutyric acid (BHB) or acetone. Chemical formulas of acetoacetate, BHB and acetone are shown in Figure (2).

Figure (2): Chemical formula of acetoacetate, BHB, and acetone

acetoacetate   BHB   acetone

The major reactions starting from the production of acetoacetate from hepatic acetyl CoA (a coenzyme present in all living cells that functions as an acyl group carrier and is necessary for fatty acid synthesis and oxidation, pyruvate oxidation, and other acetylation reactions, and assists in transferring fatty acids from the cytoplasm to mitochondria) are outlined as shown in the following equation:

$$\text{acetyl CoA} \underset{}{\overset{NAD/NADH}{\rightleftharpoons}} \text{acetoacetic acid} \underset{}{\overset{NAD/NADH}{\rightleftharpoons}} \text{beta-hydroxybutyric acid}$$

Acetyl-CoA is an important molecule in metabolism, used in many biochemical reactions. Its main use is to convey the carbon atom within the acetyl group ($C_2H_3O^+$) to the citric acid cycle to be oxidized for energy production.

There is one H^+ produced for each acid anion produced. So, there are 2 H^+ produced in the transformation from CoA to acetoacetic acid to beta-hydroxybutyric acid. This will increase the acidity from pKa 3.58 to pKa 4.70. Production of more hydrogen will also cause the HCO_3 to be more anionic (HCO_3^-), and thus more radicals (ions) will appear in the blood.

The reduction of acetyl CoA to acetoacetate to beta-hydroxybutyrate can occur in sever ethanol toxity and can increasing the acidity of the blood.

Some acetoacetate spontaneously decarboxylates to yield acetone. The odor of acetone can be smelled on the breath of individuals with severe ketosis. High levels of acetone can cause death, coma, unconsciousness, seizures, and respiratory distress. It can damage your kidneys and the skin in your mouth.

3-2 Oxidative stress

Reactive oxygen species (ROS) such as O_2^-, H_2O_2 or OH^- can damage DNA, RNA, and proteins which theoretically contribute to the physiology of aging. All small molecules with ions that include oxygen ions, free radicals and peroxides are highly reactive, due to the presence of unpaired valence shell electrons. For example, O^+ has 5 electrons in a valence shell, so there is 1 unpaired electron. F (fluorine) has 7 electrons in a valence shell, so there is1 unpaired electron. O^- has 7 electrons in valence with 1 unpaired electron. Ar

221

argon) has 8 electrons in a valence and no unpaired electrons. O (oxtgen) has 6 electrons in a valence and has no unpaired electrons and so on.

During times of environmental stress (e.g. UV, heat exposure and radiation) ROS levels can increase dramatically, which can result in significant damage to cell structures. This situation is known as oxidative stress. Oxidative stress is produced in cells by oxygen-derived species resulting from cellular metabolism and from interaction with cells of exogenous sources such as carcinogenic compounds, redox-cycling drugs and ionizing radiations. Oxidative stress can also cause oxidative damage to lipids, proteins, and DNA within cells, leading to such events as DNA strand breakage and disruption of calcium ion and minerals metabolism.

Oxidative stress can produce a multiplicity of modifications in DNA including base and sugar lesions, strand breaks, DNA-protein cross-links and base-free sites. Thus, oxidative DNA damage could be the cause of carcinogenesis and neurodegenerative diseases such as Alzheimer's disease. There is also strong evidence for the role of this type of DNA damage in the aging process. The accumulation of oxidative DNA damage in non-dividing cells is thought to contribute to age-associated diseases.

Oxidative stress can result from exposure to toxic agents. Toxic agents are generally three types of toxic entities; chemical, biological, and physical:

- Chemicals include inorganic substances such as asbestos, mercury, lead, hydrofluoric acid, chlorine and chlorine gas, organic compounds such as alcohols (methyl, ethyl, isopropyl, propyl, butyl, etc.), most medications, and poisons from living things. Ozone, oxides of nitrogen, and cigarette smoke can cause oxidative damage.

- Biological toxic entities include those bacteria and viruses that are able to induce disease in living organisms.

- Physical toxins include things not usually thought of under the heading of "toxic" by many people: These are direct blows, a concussion, sound and vibration, heat and cold, non-ionizing electromagnetic radiation, such as infrared and visible light, and ionizing radiation such as X-rays and alpha, beta and Gamma radiation. Cosmic phenomena such as sun corona, solar wind, and nuclear astrophysics fusion can also generate oxidative stress.

The balance of oxygen depends on radicals. For example, one oxygen stable molecule can be affected by one anion of radials to produce radical oxygen that can be combined with the hydrogen base in DNA and consequently damage the DNA. The damaged DNA produces hydro peroxide and oxygen. The hydro peroxide will react with the radical to produce radical hydroxide and water as per the following equation:

$$O_2 + e^- \longrightarrow O_2^-$$

$$2O_2^- + 2H^+ \longrightarrow H_2O_2 + O_2$$

$$H_2O_2 + e^- \longrightarrow OH + OH^-$$

$$OH + H^+ + e^- \longrightarrow H_2O$$

Superoxide is generated from the electron transport chain as well as the non electron transport mechanism. The major source of superoxide from the electron transport chain is radicals. Electrons generated in the oxidation of carbohydrates, proteins, and lipids reduce 1 to 4 percent of oxygen to superoxide within mitochondria*. The leakage of electrons to oxygen also takes place in the chloroplast and the endoplasmic reticulum. Superoxide is also produced during the respiratory burst of phagocytic cells (e.g. monocytes, macrophage, neutrophils and eosinophils).

* Boveris, A., and Cadenas, E. Production of superoxide radicals and hydrogen peroxide in mitochondria, *superoxide dismutase*. Vol.II,Oberly,L.W.ED., CRC Press, PP 15-30, 1982

3-3 Superoxide Dismutases (SOD)

They are a class of enzymes that help the transformation (dismutation) of the superoxide into oxygen and hydrogen peroxide. SODs are an important antioxidant shield in nearly all cells exposed to oxygen.

The SOD-catalyzed dismutation of superoxide can be written with the following half-reactions:

superoxide
anion radical

$$Fe_3^+ + O_2^- \xrightarrow{SOD} Fe_2^+ + O_2$$

$$Fe_2^+ + O_2^- + 2H^+ \xrightarrow{SOD} Fe_3^+ + H_2O_2$$

$$Cu^{2+} + O_2^- \xrightarrow{SOD} Cu^{1+} + O_2$$

$$Cu^{1+} + O_2^- + 2H^+ \xrightarrow{SOD} Cu^{2+} + H_2O_2$$

There are three main groups of superoxide dismutase:

• Copper and zinc SOD which is most commonly used by species of large organisms such as animals, plants and fungus (eukaryotes). The group is called Cu-Zn-SOD.

- Iron or manganese SOD which is used by prokaryotes (group of organisms that have nocell nucleus) such as E-coli bacteria.

- Nickel SOD which is used by prokaryotes.

In humans, there are three types of SODs:

1. SOD1 is a copper and zinc containing enzyme located in the cytoplasm
2. SOD2 is a manganese containing enzyme in the mitochondria
3. SOD3 is a copper and zinc in the extracellular area.

SOD is an enzyme that inactivates excess free radicals, preventing them from damaging cell membranes. SOD is taken by some athletes as an ergogenic aid to protect their bodies against the many free radicals produced during vigorous exercise. However, when taken orally the protein of the enzyme is digested and made useless. It is an enzyme widely distributed in the body, which destroys superoxide (O_2^-) radicals released during aerobic metabolism. These free radicals have been linked with cancer and degenerative diseases.

3-4 Anabolic hormones

Anabolic means "building up" and catabolic means "breaking down". In other words, Anabolic means your body is in growth mode, you are getting bigger. Catabolic means your body is in decrease mode, suffering muscle wastage, you are getting smaller. Anabolic is endergonic which means building up muscles and organic cells in which more energy is stored in the products than in the reactants. Catabolic is exergonic which means cellular respiration in which less energy is stored in the products than in the reactants, Figure (3).

Figure (3): Anabolic and catabolic energy

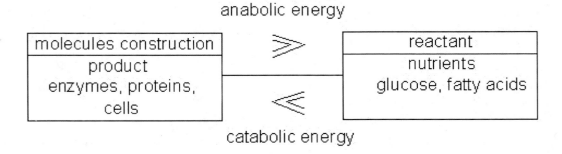

By reducing the rate of catabolism, anabolism increase, resulting in faster growth rate and a higher performance rate. Athletes use nutrients such as L-Gluthamine to reduce the effect of cortisol which is increased at a high level when the muscles are strenuously exercised. Cortisol speeds up the breakdown of tissues and increases the rate of catabolism.

Aging is associated with diseases that affect the bones, such as osteoporosis and primary hyperparathyroidism, which result in gradual net loss of bone mineral (such as calcium, magnesium, manganese, zinc, copper and boron). This leads to osteopenia (low bone mineral density) which is a leading cause of fractures in adults. Anabolic hormones are key regulators of bone mass and osteopenia. They may result from reduction in anabolic hormones, such as loss of estrogen production in post-menopausal women or of androgens in older men. Also with excessive production of parathyroid hormone (PTH) as in primary hyperparathyroidism, or glucocorticoid* excess as a consequence of large doses of steroid use in immunosuppressive therapy, for example, with organ transplantation, allergies, and asthma. Other imbalances in local growth factors and/or bone-active cytokines (proteins that are produced by cells, and interact with cells of the immune system in order to regulate the body's response to disease and infection) result from a variety of conditions (e.g. chronic inflammation, rheumatoid arthritis) and may also contribute to osteopenia.

* Glucocorticoids (GC) are steroid hormones that bind to the glucocorticoid, which is present in almost every human cell. Glucocorticoid derives from the role in the regulation of the metabolism of glucose, the synthesis in the adrenal cortex, and their steroidal structure.

There are mainly eight patient anabolic hormones. Decreases in these hormones have been hypothesized to contribute to the loss of muscle and bone mass, as well as cognitive and immune function, in aging adults:

- Growth hormone
- IGF-1 (Insulin-like growth factors)
- Insulin
- Somatostatin
- Testesteone
- Estradiol and estrogen
- Orexin / hypocretin
- Melatonin

3-4-1 Growth Hormone (GH)

The growth hormone (GH) is a protein hormone which consists of a 191-amino acid, single-chain protein polypeptide hormone that is synthesized, stored, and secreted by the somatotroph cells in response to somatocrinin hormone received from the hypothalamus within the lateral wings of the anterior pituitary gland, Figure (4). In children, GH has growth-promoting effects on the body. It stimulates the secretion of somatomedins from the liver, which are a family of insulin-like growth factor (IGF) hormones. These, along with GH and thyroid hormone, stimulate linear skeletal growth in children. In adults, GH stimulates protein synthesis in muscle and the release of fatty acids from adipose tissue (anabolic effects). It inhibits uptake of glucose by muscle while stimulating uptake of amino acids. The amino acids are used in the synthesis of proteins, and the muscle shifts to using fatty acids as a source of energy. GH secretion occurs in a pulsatile (short, concentrated secretion) and sporadic manner. Thus, a single test of the GH level is usually not performed. Because of the irregular release of GH, the patient will have his blood drawn a total of five times over a few hours.

Figure (4): Pathways of the growth hormone

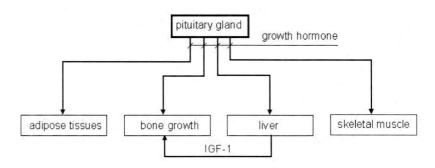

adipose tissue or body fat is composed of adipocytes. Its main role is to store energy in the form of fat, it also cushions and insulates the body. Obesity and overweight do not depend on body weight but on the amount of adipose fat.

Aging is affected by the growth hormone which is modulated by many factors, including stress, exercise, nutrition, sleep and the growth hormone itself. However, its primary controllers are:

- Hypothalamic growth hormone-releasing hormone (GHRH)
- Hypothalamic somatostatin horomone (SS)
- Stomach hormone (Ghrelin)

All of above hormones stimulate both the synthesis and secretion of growth hormone. The modulation is done through feed back control adjustment, Figure (5). For example, if the GHRH is increased, the pituitary gland sends a signal to the hypothalamus gland ordering it to reduce the somatostatin hormone.

Figure (5): Modulation of growth hormone

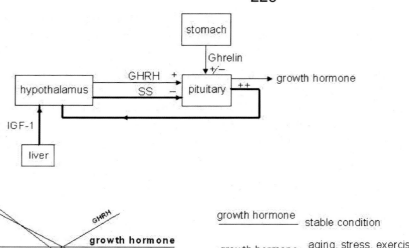

Growth hormone secretion is also part of a negative feedback loop involving IGF-I. High IGF-I in blood level leads to decreased secretion of the growth hormone by directly suppressing the release of somatostatin from the hypothalamus. Generally, the growth hormone is inversely proportional with age, Figure (6).

Figure (6): Growth hormone and age

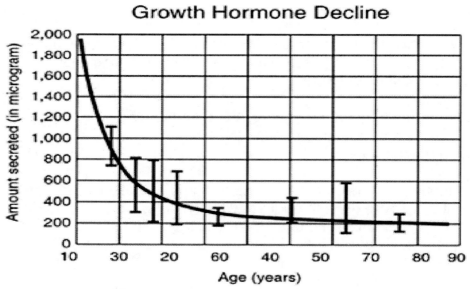

Growth Hormone (GH) levels in the blood decline about 14% per decade after age 25 due to decreased GH releasing hormone. Additionally, much as insulin resistance increases with age, GH receptors become less responsive to GH with aging ("GH resistance"). GH decline is blamed for age-related adiposity as well as loss of muscle mass & bone mineral.

In children and young adults, the most intense period of growth hormone release is shortly after the onset of deep sleep.

Clinically, deficiency in the growth hormone or defects in its binding to the receptor are seen as growth retardation, dwarfism, or aging. The symptoms of growth hormone deficiency depend upon the age of the onset of the disorder and can result from either heritable or acquired disease. On the other hand, the effect of excessive secretion of the growth hormone is seen as two distinctive disorders:

- Giantism (or gigantism) is the result of an excessive growth hormone secretion that begins in young children or adolescents.
- Acromegaly results from an excessive secretion of growth hormone in adults, usually the result of benign pituitary tumors, Figure (7).

Figure (7): Excessive secretion of growth hormone

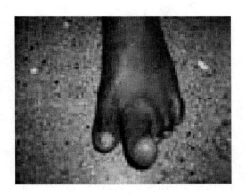

3-4-1-1 Uses of growth hormone

- In the past, growth hormones extracted from human cadaver pituitaries were used to treat children with growth retardation. Recently, growth hormones are produced using recombinant DNA technology.

- Growth hormones can be extended to treatment of normal children - so called "enhancement therapy". Similarly, growth hormones have been used by some to enhance athletic performance.
- Cosmetic therapy, using growth hormones is used for enhancement of aged skin and colour.

Although growth hormone therapy is generally safe, it could entail unpredictable health risks. Parents that request growth hormone therapy for their children of normal stature are clearly misguided.

3-4-2 Insulin/Hormone IGF-1

Insulin-like growth factor 1 (IGF-1) is a polypeptide protein hormone (C_{331}-H_{512}-N_{94}-O_{101}-S_7) similar in chemical bonding to insulin ($C_{256}H_{381}N_{65}O_{79}S_6$). It plays an important role in childhood growth and continues to have anabolic effects in aged people. In the 1970's, IGF-1 was known as as "Sulfation Factor" and "Nonsuppressible Insulin-like Activity" (NSILA). In the 1980's, it was known as "Somatomedin C". The most popular type of IGF-1 available on the Black Market is known as Long R3 Insulin-like Growth Factor-I or Long R3IGF-I. LR3igf-1 is more potent than the lesser versions which are no longer available on the Black Market.

IGF-1 is released in the liver and binds to the IGF receptors in the cells, which ultimately causes a stimulation of cell growth (both causing new tissue formation and tissue growth) and an inhibition of cell death (apoptosis). It is a highly anabolic and anti-catabolic compound. For the athlete or bodybuilder, this had many positive effects: increased nitrogen retention (one molecule has 94 atoms of nitrogen) and protein synthesis because it is highly anabolic. IGF-1 promotes the growth of new muscle cells and connective tissue cells, and improves collagen formation and cartilage repair. IGF protects the neurons of the brain as well as promotes growth of new motor neurons, making it more possible to rapidly learn new skills. It also enhances the strength of bones through bone production and repair. Many of the effects of GH are due to IGF–1, which is produced in the liver as a result of GH stimulation and is also produced in exercised skeletal muscle to increase growth of the muscle. But although exercise can increase GH and IGF–1 concentrations in healthy young athletes, no such effect is seen for older people (who benefit, nonetheless, from increased aerobic capacity, muscle strength and fat-free mass) [Journal of Gerontology;Vitiello, M;52A(3):M149-M154 (1997)].

Men and women taking GH replacement therapy showed lower serum LDL cholesterol (Cholesterol travels through the blood in different types of packages called lipoproteins. Low-density lipoproteins (LDL) deliver cholesterol to the body, and high-density lipoproteins (HDL) remove cholesterol from the bloodstream). and lower LDL/HDL ratios, reduced body fat and increased lean mass [The Journal of Clinical Endocrinology & Metabolism; Hoffman, AR; 89(5): 2048-2956

(2004)]]. IGF–1 improvement in vascular density can reduce age-associated ischema (a shortage of oxygen, glucose and other blood-borne fuels) in the heart & brain [Journal of Anti-Aging medicine; Sonntag, WE; 4(4):311-329 (2001)]. GH replacement significantly reduces plasma inflammatory cytokines (IL-6 & TNF-alpha) [The Journal of Clinical Endocrinology & metabolism; Serri,O ;84(1):58-63 (1999)].

IGF–1 appears to stimulate the activity of Natural Killer (NK) cells. IGF–1 decline may cause much of the decline in T-lymphocyte activity with aging In fact, GH/IGF–1 replacement can reverse the age-related involution of the thymus gland.

IGF-1 tests provide an accurate assessment of the growth hormone levels. An IGF-1 outside the normal range can serve as a starting point for correcting the cause contributing to premature aging, Table (1).

Table (1): Corresponding IGF-1, GH and age of male and female (ng means nanogram, 10^{-9}, mc means microgram, 10^{-6}).

Age	Male ng/ml	Female ng/ml	GH Secretion
2 to 5 yrs.	17 - 248	17 - 248	?
6 to 8 yrs.	88 - 474	88 - 474	?
9 to 11 yrs.	110 - 565	117 - 771	?
12 to 15 yrs.	202 - 957	261 - 1096	2000 mcgs.
16 to 24 yrs.	182 - 780	182 - 780	500 mcgs.
25 to 39 yrs.	114 - 492	114 - 492	250 mcgs.
40 to 54 yrs.	90 - 360	90 - 360	125 mcgs.
55 + yrs.	71 - 290	71 - 290	60 mcgs
80 yrs.	1 - 71	1 - 71	30 mcgs

Men and women taking GH replacement therapy showed lower serum LDL cholesterol and lower LDL/HDL ratios, reduced body fat and increased lean mass*. Although blood lipid levels improve, long term mortality benefit have not yet been demonstrated. Treatment of healthy, normally aging individuals found the only benefit to be a slight increase in muscle mass, with frequent side-effects and no evidence that it is safe to use long-term. IGF–1 improvement in vascular density can reduce age-associated ischemia** in the heart & brain. GH replacement significantly reduced plasma inflammatory cytokines (IL-6 & TNF-alpha.

230

*The Journal of Clinical Endocronology& Metabolism;Hoffman,AR; 89(5):2048-2056 (2004)

** Ischemia is the lack of delivery of oxygen to a major organ, most often affecting the heart or the brain. Blood flow is either blocked, or the blood flowing to the organ has extremely low oxygen saturation. Since all of the body's tissues need oxygen to maintain function, ischemia can result in shut down of an organ or significant damage to the organ.

On January 23, 1998 researchers at the Harvard Medical School released a major study providing conclusive evidence that IGF-1 is a potent risk factor for prostate cancer; see Chemistry, Biology and Cancer: The bond by the author of this book. Should you be concerned? Yes, you certainly should, particularly if you drink milk produced in the United States, see http://www.vvv.com/healthnews/milk.html

All mammals produce IGF-1 hormones very similar in structure and human and bovine IGF-1 are completely identical. IGF-1 is known to promote the growth of both normal and cancerous cells. In 1990 researchers at Stanford University reported that IGF-1 promotes the growth of prostate cells. It was also discovered that IGF-1 accelerates the growth of breast cancer cells. In 1995 researchers at the National Institutes of Health reported that IGF-1 plays a central role in the progression of many childhood cancers and in the growth of tumors in breast cancer, small cell lung cancer, melanoma, and cancers of the pancreas and prostate. In September 1997 an international team of researchers reported the first epidemiological evidence that high IGF-1 concentrations are closely linked to an increased risk of prostate cancer. Other researchers provided evidence of IGF-1's link to breast and colon cancers. Bovine growth hormone was first synthesized in the early 1980s using genetic engineering techniques (recombinant DNA biotechnology, see recombinant DNA in this book). Small-scale industry-sponsored trials showed that it was effective in increasing milk yields by an average of 14 per cent if injected into cows every two weeks. In 1985 the Food and Drug Administration (FDA) in the United States approved the sale of milk from cows treated with rBGH, recombinant bovine growth hormone, (also known as BST, bovine somatotropin) in large-scale veterinary trials. In 1993 they approved commercial sale of milk from rBGH-injected cows. At the same time the FDA prohibited the special labeling of the milk so as to make it impossible for the consumer to decide whether or not to purchase it. Furthermore, recent research has shown conclusively that the levels of a hormone called "insulin-like growth factor-1" (IFG-1) are elevated in dairy products produced from cows treated with rBGH. Canadian and European regulators have found that the FDA completely failed to consider a study that showed how the increased IGF-1 in rBGH milk could survive digestion and make its way into the intestines and blood stream of consumers. These findings are significant because numerous studies now demonstrate that IGF-1 is an important factor in the growth of cancers of the breast, prostate and colon.

3-4-2-1 Insulin

In 1921, Frederick Grant Banting and Charles H. Best from the university of Toronto, Canada, discovered insulin. Banting and Best extracted material from the pancreas of dogs. They first used this material to keep diabetic dogs alive and in 1922 they used it successfully on a 14-year-old boy with diabetes. In 1923, Banting and Macleod were awarded the Nobel Prize. Best and Collip were overlooked but Banting and Macleod shared the prize money with them.

Insulin is a natural hormone made by the pancreas that controls the level of the sugar glucose in the blood. Insulin permits cells to absorb glucose for energy. Cells cannot absorb glucose without insulin. Insulin is made by the beta cells in the islets of Langerhans in the pancreas. If the beta cells are damaged or degenerate, the sugar in the body increases and type I diabetes is the result. If the cells throughout the body resist or do not respond to the insulin produced by the beta cells, type II diabetes is the result.

Insulin is a hormone best known for its role in glucose metabolism. Insulin is a peptide (peptide hormones are synthesized from amino acid according to an mRNA template, which is itself synthesized from a DNA template inside the cell's nucleus) composed of 51 amino acids and has a molecular weight of 5808 Da (Dalton is the approximate mass of a hydrogen atom, a proton, or a neuron and equals $1.660538782(83) \times 10^{-27}$ kg).

Insulin causes cells to take glucose from the blood to be distributed in the body's organs as shown in Figure (8).

Figure (8): Glucose distribution after a meal

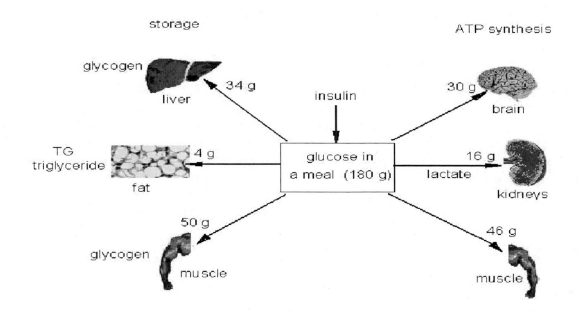

The brain consumes 30 grams out of 180 grams of glucose (about 16%). The largest consuming organs are the muscles followed by the liver. If the insulin is reduced, the sugar to the brain will be constant because the insulin modulates the glycogen. So, after a meal insulin secretion is activated, glucagon secretion is minimized, and the liver takes up glucose which is then stored as glycogen to be used to buffer blood glucose at a later time. Insulin also stimulates glucose uptake and glycogen synthesis in muscles. Note that muscle glycogen cannot be released to the circulation. Muscle glycogen is used exclusively as a substrate for muscle activity. Insulin has many functions in fatty acids, muscle tone, electrolytes, pH level, esterification, break down of protein, degradation of damaged cells, etc, Figure (9).

Figure (9): Effect of Insulin on metabolism

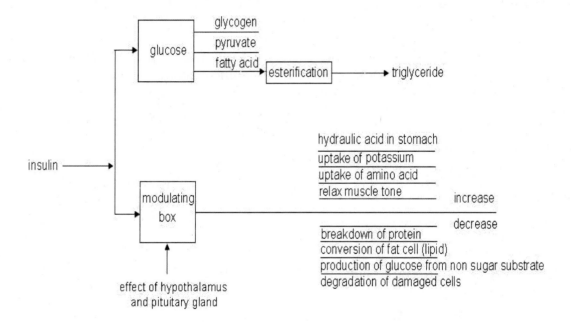

Insulin controls the acetyl CoA which is the central role for all classes of biomolecules and the major source of useful metabolic energy. Figure (10) summarizes all metabolisms and the central role of acetyl CoA. Note that acetyl CoA can react reversibly. The interactions of amino acids with acetyl CoA and the citric acid cycle will be studied in protein metabolism. Notice that acetyl CoA can react "reversibly" in the break down or synthesis of lipids and amino acids. This is not the case with carbohydrate (glucose, fructose, and galactose) metabolism. In mammals, acetyl CoA cannot make carbohydrates.

Figure (10): Role of CoA in metabolism

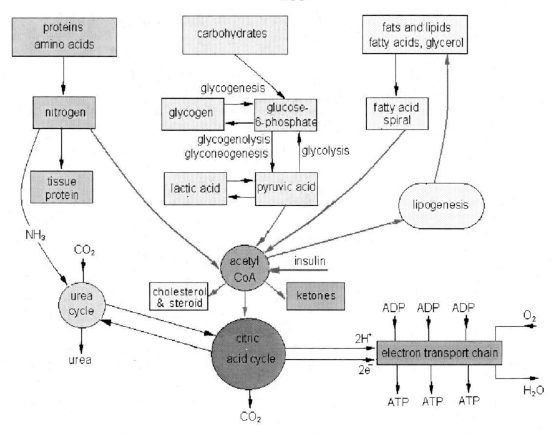

During fasting, insulin drops and the cycle will be reversed, i.e. ketones, cholesterol, fats and fatty acid will convert to CoA, and fat will convert to glycerol to glucose as shown in Figure (11).

Figure (11): Reversed cycle during fasting

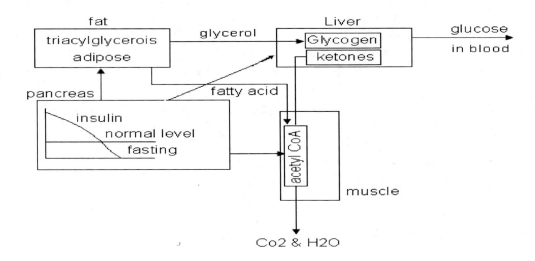

During fasting, or when insulin is below normal level, the following diabetic metabolism occurs:

- decreased cellular glucose uptake which results in hyperglycemia, glucosuria, osmotic diuresis, and electrolyte depletion
- increased catabolism of protein which results in increased plasma amino acid and nitrogen in the urine
- increased lipolysis which results in increased plasma FFA, ketogenesis, ketoria, and ketonemia
- dehydration acidosis
- Auditory and visual information process more slowly when the brain is temporarily deprived of its sugar supply. The brain does not store sugar, it depends on the sugar of the flowing blood
- Decreased cortisol in the first two days of starvation, and then it increases slightly
- Coma and death

3-4-2-2 Insulin and aging

Human cells are insulin resistant because insulin has a toxic effect. Different cells respond to insulin differently. Some cells are more resistant than others, as some cells are incapable of becoming very resistant. The liver becomes resistant first, followed by the muscle tissue and lastly the fat cells. "Insulin resistance syndrome" refers to risk factor for type 2 diabetes, low HDL ("good") cholesterol, abdominal obesity and high blood pressure. Symptoms of insulin resistance include:

- Increased weight and fat storage. For most people, too much fat is too much weight. In males, a large abdomen is the more obvious and earliest sign of Insulin Resistance, in females, it's prominent buttocks. Overweight is usually associated with increased triglycerides which are stored in the arteries and blood vessels.
- Too much insulin due to insulin resistance can cause increased blood pressure. There is a direct relation between the level of insulin and blood pressure.
- Insulin resistance can cause physical and mental fatigue (brain fogginess). Brain fogginess is associated with poor memory, loss of creativity and learning disabilities.
- Insulin Resistance sufferers who eat lots of carbohydrates suffer from intestinal bloating (gas) which makes them get sleepy, or feel agitated, jittery and moody.
- Fatigue is the most common feature of insulin resistance; it makes people tired quickly. Some are tired just in the morning or afternoon, others are exhausted all day.
- Because carbohydrates are a natural "downer" depressing the brain, it is common to see many depressed people who also have insulin resistance.

Professor Marc Tatar at Brown University said "insulin is a direct player in the aging process". Tatar's research is part of a growing body of evidence linking low insulin levels to increased longevity. In recent years, scientists have found that mice and other animals live longer when they eat a low-calorie diet, which reduces insulin production. Brown University scientists found that insulin:

- Plays a central role in aging. According to the study, when the chemical messages sent by an insulin-like hormone are reduced inside the fat cells of a fruit fly, the fly's lifespan increases significantly, by an average of 50 percent. This is important, since fruit flies and humans share some 13,601 genes.
- Regulates its own synthesis. The Brown University experiments shed important light on the role insulin plays in the regulation of its own synthesis. The study shows that if you block the hormone's action inside a few specific cells, the entire body stays healthier longer.

3-4-2-3 Tips to reduce insulin:

- Yasuyuki Nakamura, associate professor of internal medicine at Shiga University of Medical Science, said the study showed men could reduce the risk of death from conditions such as heart disease or stroke by some 30 percent by eating fish once every two days.
- Cut back on refined carbohydrates. Refined carbohydrates are probably the top controllable aging factor. Diets should include little refined grains and sugar, and should have more complex hydrocarborate found in most vegetables, protein and natural fats. Avoid refined and hydrogenated oils. Diets should be moderate in fresh fruits, but processed fruits as found in pastries, jellies and bottled juices are not acceptable. Have fresh vegetable juice.
- Fasting could increase longevity. Studies have shown that when rodents are fed all they can eat, but fasted every two, three, or four days, they have an increase in longevity.
- Use a full spectrum antioxidant formula. This will minimize the destructive aging impact of free radicals produced in the body.
- Use an L-carnosine based formula. This minimizes the impact of advanced glycation end products, which can destroy proteins, break down organs, and stiffen connective tissue. Glycation is the result of a sugar molecule, such as fructose or glucose, bonding to a protein or lipid molecule Researchers in Britain[1], South Korea, Russia and other countries have shown that carnosine has a number of antioxidant properties that may be beneficial. Carnosine has been proven to scavenge reactive oxygen species as well as alpha-beta unsaturated aldehyde formed from peroxidation of cell membrane fatty acid during oxidative stress.
- Aging is an independent risk factor for hypertension, and hypertension and insulin resistance commonly coexist in the elderly. Hypertension is related to an increased level of insulin. Hypertension can be reduced by weight

loss, limiting smoking, salt and alcohol, eating fruit and vegetables containing potasium (banana), and having regular exercise.

3-4-2-4 Gene daf-2

This gene has insulin-like metabolic action. A team at the University of Missouri prepared SAGE (serial analysis of gene expression) libraries from roundworms that contained a mutated form of the daf-2 gene (insulin receptor-like protein), which is a principal lifespan-determining factor in C. elegans. Worms that lack fully functional daf-2 exhibit significantly extended lives, persisting approximately twice as long as their wild-type counterparts.

Decreased DAF-2 signaling also causes an increase in life-span. Life-span regulation by insulin-like metabolic control is analogous to mammalian longevity enhancement induced by caloric restriction, suggesting a general link between metabolism, diapause (slow-down of metabolism), and longevity. This gene family is a part of major metabolic pathways including lipid, protein, energy metabolism, stress response, and cell structure. Similar expression patterns of closely related family members emphasize the importance of these genes in the aging-related processes.

3-4-3 Somatostatin

Somatostatin is a polypeptide hormone produced mostly by the hypothalamus that inhibits the secretion of various other hormones, such as growth hormone, somatotropin, glucagon, insulin, thyrotropin, and gastrin. Two forms of somatostatin are synthesized. They are referred to as SS-14 and SS-28, reflecting their amino acid chain length. The relative amounts of SS-14 versus SS-28 secreted depend upon the tissue. For example, SS-14 is the predominant form produced in the nervous system and apparently the sole form secreted from the pancreas, whereas the intestine secretes mostly SS-28. SS-28 is roughly ten-fold more potent in inhibition of growth hormone secretion, but less potent than SS-14 in inhibiting glucagon release. Generally, somatostatin inhibits the secretion of many other hormones such as gastrointestinal hormones (gastrin, motelin, vasoactiveintestinal peptide, gastric inhibitorypolypeptide, cholecystokininin ,etc), pancreatic hormones (insulin and glucagon), and exocrine secretory action of the pancreas.

Somatostatin affects neuromodulatory activity within the central nervous sytem, and appears to have a variety of complex effects on neural transmission. Injection of somatostatin into the brain of rodents leads to an increased arousal and decreased sleep, and impairment of some motor responses.

Somatostatin and its synthetic analogs are used clinically to treat a variety of neoplasms. It is also used to treat giantism and acromegaly (extremities enlargement), due to its ability to inhibit growth hormone secretion.

237

The interaction of somatostatin with body organisms is illustrated in Figure (12).

Figure (12): Interaction of somatostatin with organisms

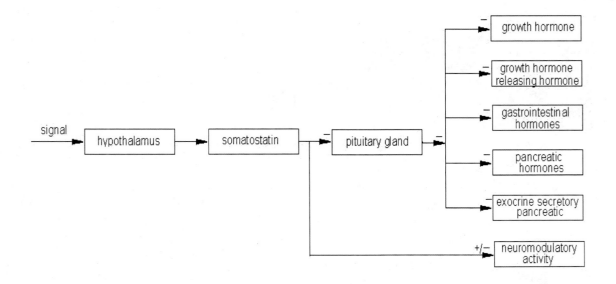

3-4-3-1 Somatostatin and aging

There were no significant age-related changes in somatostatin levels in the human frontal cortex, caudate nucleus, putamen, medial globus pallidus, or substantia nigra as cited by http://linkinghub.elsevier.com/retrieve/pii/.
The following patern shows the relationship between somatostatin and the growth hormone, Figure (13).

Figure (13): Relationship between somatostatin and the growth hormone in young, adult, and centenarian people.

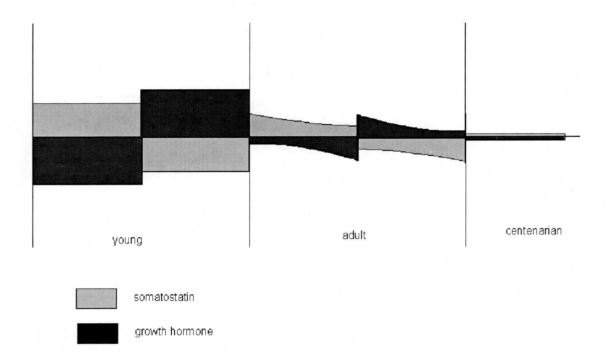

somatostatin

growth hormone

3-4-4 Testosterone

The testes produce testosterone regulated by a complex chain of signals that occurr in the brain. Signals start from the hypothalamic to pituitary to gonadal membranes. The hypothalamus secretes a gonadotropin-releasing hormone (GnRH) to the pituitary gland in pulses (bursts), which triggers the secretion of the leutenizing hormone (LH) from the pituitary gland. The leutenizing hormone stimulates the Leydig cells of the testes to produce testosterone. Normally, the testes produce 4–7 milligrams (mg) of testosterone daily in adults. Normal Results:

- Male: 300 -1,000 ng/dL
- Female: 20 - 80 ng/dL

In females the process is reversed and the testosterones converted to estrogen in the hypothalamus and the pituitary gland, Figure (14).

Figure (14): testosterone in male and female

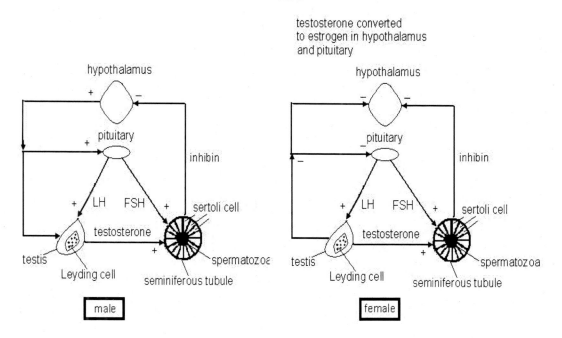

Follicle stimulating hormone (FSH) is a hormone synthesized and secreted by gonad tropes in the anterior pituitary gland. FSH and LH act synergistically in reproduction.

In men, FSH enhances the production of androgen-binding protein by the Sertoli cells of the testes and is critical for spermatogenesis.
In women, in the ovary FSH stimulates the growth of immature Graafian follicles to maturation. As the follicle grows it releases inhibin, which shuts off the FSH production.

Testosterone is a steroid hormone from the androgen group. Testosterone is mainly secreted in the testes of males and the ovaries of females. Small amounts are also secreted by the adrenal glands. It is mainly a male sex hormone. Testosterone production declines naturally with age. Low testosterone, or testosterone deficiency, may result from disease or damage to the hypothalamus, pituitary gland, or testicles. This inhibits hormone secretion and testosterone production, and is also known as hypogonadism. Metabolism of testosterone occurs in the liver, the testis, and the prostate. Depending on age, insufficient testosterone production can lead to abnormalities in muscle and bone development (osteoporosis), underdeveloped genitalia, and diminished virility. On average, an adult human male body produces about forty to sixty times more testosterone than an adult human female body. One can see chemical formulas are almost the same for testosterone, androgen, progesterone and steroid, Figure (15).

Figure (15): Chemical formulas of main androgen groups

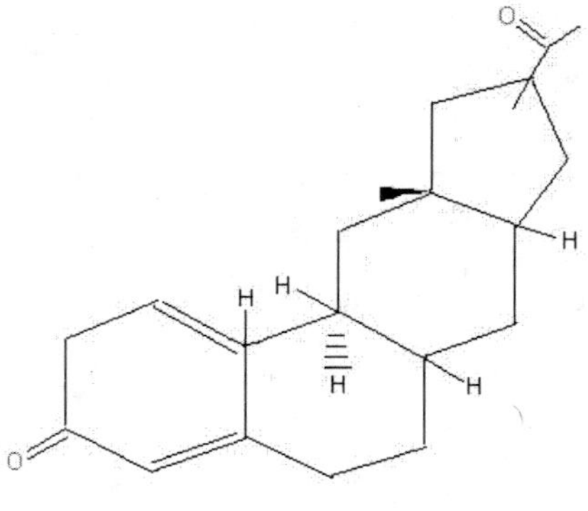

androgen/progesterone

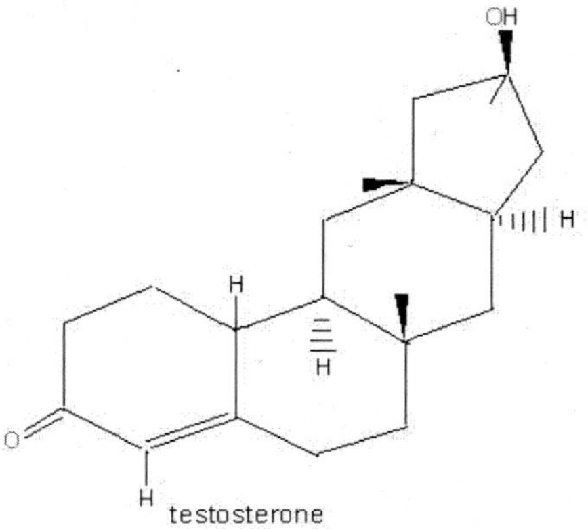

testosterone

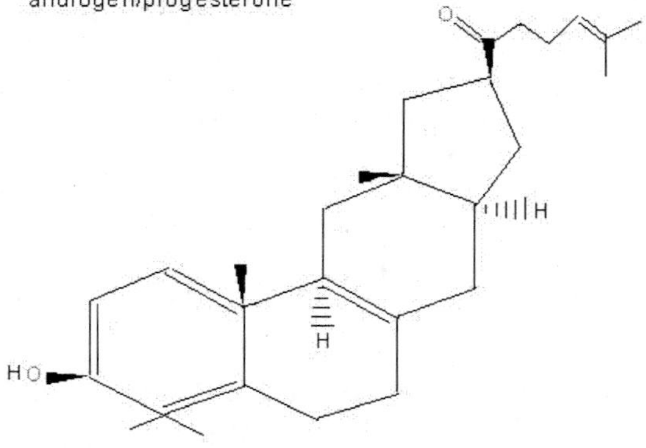

steroid

3-4-4-1 Increased production of testosterone:

- Androgen resistance
- Congenital adrenal hyperplasia, which can affect both boys and girls. People with congenital adrenal hyperplasia lack an enzyme needed by the adrenal gland to make the hormones cortisol and aldosterone.
- Ovarian cancer which is a cancerous growth arising from different parts of the ovary. The most common ovarian cancer, more than 80%, arises from the outer lining of the ovary. Other forms of ovarian cancer arise from the egg cells which are referred to as germ cell tumor.
- Testicular cancer. The exact cause of testicular cancer is unknown. There is no link between vasectomy (which is a surgery to cut the vas deferens, the tubes that carry a man's sperm from his scrotum to his urethra. The urethra is the tube that carries sperm and urine out of the penis. After a vasectomy, sperm cannot move out of the testes. A man who has had a successful vasectomy cannot make a woman pregnant) and testicular cancer. Factors that may increase a man's risk for testicular cancer include:
  - Abnormal testicle development
  - History of testicular cancer
  - History of undescended tisticles(s)
  - Klinefelter syndrome (which is the presence of an extra X chromosome in a male).

- Precocious puberty, which is the time during which sexual and physical characteristics mature. Precocious puberty is when these body changes happen earlier than normal.
- Polycystic ovary disease, which is a condition in which there are many small cysts in the ovaries, which can affect a woman's ability to get pregnant.

3-4-4-2   Decreased Production of Testosterone

This can cause:

- Chronic illness
- Delayed puberty
- Testecular failure, which is the inability of the testicles to produce sperm or male hormones).
- Prolactinoma, which is a prolactinoma, a noncancerous pituitary that produces a hormone called prolactin. This results in too much prolactin in the blood. Prolactin is a hormone that triggers lactation or milk production. Prolactin stimulates the mammary glands to produce milk (lactation). Increased serum concentrations of prolactin during pregnancy cause enlargement of the mammary glands of the breasts and increases the production of milk.
- Hypopituitarism, which is a condition in which the pituitary gland does not produce normal amounts of some or all of its hormones.

3-4-5 Estradiol and estrogen

Like testosterone is a form of an androgen group, estradiol is a form of estrogen. Estrogen is manufactured mostly in the ovaries, by developing egg follicles. In addition, estrogen is produced by the corpus luteum in the ovary, as well as by the placenta. The liver, breasts and adrenal glands may also contribute to estrogen production, although in smaller quantities.

Estrogen contributes to the development of secondary sex characteristics, which are the defining differences between men and women that don't relate to the reproductive system. In women, these characteristics include breasts, a widened pelvis, and increased amounts of body fat in the buttock, thigh and hip region. Estrogen also contributes to the fact that women have less facial hair and smoother skin then men.
Estrogen is an essential part of a woman's reproductive process. It regulates the menstrual cycle and prepares the uterus for pregnancy by enriching and thickening the endometrium, Figure (16). Two hormones, the luteinizing hormone (LH) and the follicle stimulating hormone (FSH), help to control how the body produces estrogen in women who ovulate.

Figure (16): Components of corpus luteum and uterus

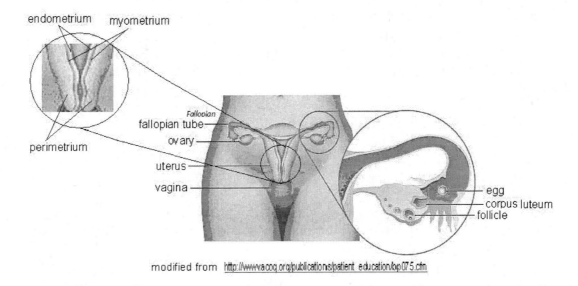

Estradiol is a formula of steroids, Figure (17). The OH in estradiol replaces the double bonded oxygen in a steroid.

Figure (17): Estradiol chemical formula

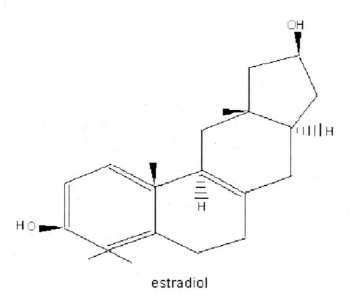

estradiol

Estradiol is actually a type of estrogen, which is the major female reproductive hormone. Estradiol is the primary type of estrogen, and it is produced in ovaries. As they grow and develop, egg follicles secrete estradiol, helping to trigger the rest of the reproductive cycle. An estradiol test is performed in order to determine a woman's ovarian reserve. It is also performed in order to confirm a woman's FSH test. It is performed on day 3 of the menstrual cycle to determine ovarian reserve. A problem with ovarian reserve results in a high level of estradiol, which can suppress the amount of FSH. Suppressed FSH can lead to lower pregnancy rate, poor ovulation, and poor success rate of IVF (in vitro fertilization). IVF is done through injectable gonadotropins (follicle stimulating hormone, FSH). FSH is one of the most important hormones involved in natural menstrual cycle, and involved in producing mature eggs. A woman can be 42 and still have some good quality eggs (and still be fertile), or she can be 25 with poor quality eggs (and be infertile), although this is rare. Figure (18) shows gonadotropin and steroid levels during the menstrual cycle.

Figure (18): Plasma gonadotropin and steroid levels during menstrual cycle

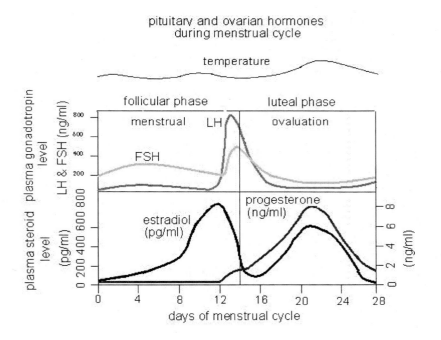

Luteinizing hormone (LH) and follicle-stimulating hormone (FSH) are called gonadotropins because thet stimulate the gonads (in males, the testes, and in females, the ovaries). They are not necessary for life, but are essential for reproduction. These two hormones are secreted from cells in the anterior pituitary called gonadotrophs. The main regulator of LH and FSH secretion is gonadotropin-releasing hormone (GnRH, also known as LH-releasing hormone). GnRH hormone regulates LH and FSH hormones in a negative feed back loop as shown in Figure (19). This regulatory loop leads to pulsatile secretion of LH and, to a much lesser extent, FSH. Pulses of secretion varies between few pulses per day to few per hour. There are at least two additional hormones secreted by the gonads. For example, hormones inhibin and activin inhibit the secretion of FSH from the pituitary.

Figure (19): LH and FSH-regulating GnRH

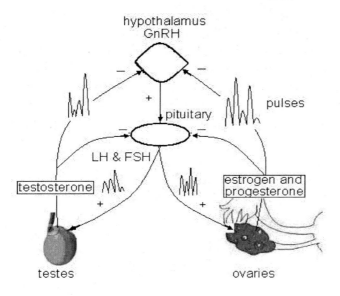

3-4-5-1 Estradiol and aging

The main circulating estrogen during the premenopausal year is 17β-estradiol which has large effect on cell growth, alkaline phosphatase (ALP) activity, and in developing connective tissue formed by a mineralized matrix that confers its elastic and strength properties of bones. Levels of this hormone are controlled by the developing follicle and resultant corpus luteum. Estradiol is reduced from 120 pg/ml to 18 pg/ml at the age of 50. The declining follicle numbers of human ovaries at age 37 is shown in Figure (20). The declining of follicle numbers and the esradiol levels are due to the changes in the endocrine system with advancing age.

Figure (20): Declining of follicle numbers with advancing age

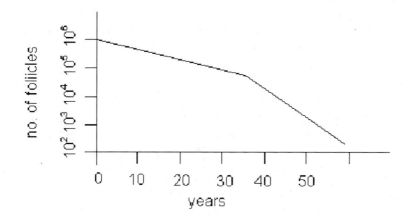

3-4-6 Orexin/Hypocretin

Orexins is also called hypocretins, and are peptide neurotransmitters expressed in neurons of the dorsal lateral hypothalamic area. Orexin regulates sleep, appetite, and energy consumption. Recent evidence indicates that it is also involved in pleasure/reward-seeking. When orexin neurons degenerate in mice, hypophagia (hypophagia induces weight loss due to reduced food intake) and obesity are observed, symptoms that are also present in some human narcoleptics. Obesity is a measurement of the body mass index (BMI) which compares a person's weight and height, though it does not actually measure the percentage of body fat. Narcoleptic is a disorder characterized by sudden and uncontrollable, though often brief, attacks of deep sleep, sometimes accompanied by paralysis and hallucinations.

The two related peptides (orexin-A and B, or hypocretin-1 and -2), with approximately 50% sequence identity, are produced by cleavage of a single precursor protein. Orexin-A/hypocretin-1 is 33 amino acid residues long and has two intrachain disulfide bonds, while Orexin-B/hypocretin-2 is a linear 28 amino acid residue peptide. Both types are produced by hypothalamic neurons and have key roles in regulating sleep and appetite.

Central administration of orexin A/hypocretin-1 strongly promotes wakefulness, increases body temperature, locomotion and elicits a strong increase in energy expenditure. A link between orexin and Alzheimer's disease has been recently suggested*. The enigmatic protein amyloid beta builds up over time in the brain and is correlated with Alzheimer's disease. Amyloid beta is a deposition of a proteinaceous mass that is converted to intraneuronal neurofibrillary tangles in the brain. Orexin has many functions in body building and on the body for life as seen in Figure (21).

Figure (21): Orexin / hypocretin and functions on the body for life

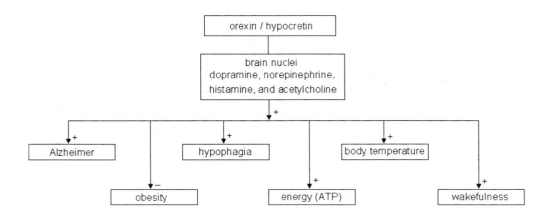

* J. E. Kang, M. M. Lim, R. J. Bateman, J. J. Lee, L. P. Smyth, J. R. Cirrito, N. Fujiki, S. Nishino and D. M. Holtzman (2009). "Amyloid-{beta} Dynamics Are

Regulated by Orexin and the Sleep-Wake Cycle". Science. Doi:10.1126/science.1180962.

3-4-7 Melatonin

Melatonin is a hormone secreted by the pineal gland in the brain that helps regulate other hormones and maintains the body's circadian rhythm. The melatonin is also produced by the retina, lens, GI (gastrointestinal) tract and other tissues. The largest organ in humans to biosynthesize melatonin is the skin. Melatonin hormone has a mechanism by which some amphibians and reptiles change the colour of their skin. Melatonin can suppress libido (sexual desire) by inhibiting secretion of luteinizing hormone (LH) and follicle stimulating hormone (FSH) from the anterior pituitary gland. It also regulates ACTH hormones that stimulate the adrenal cortex, and DHEA androgen, Figure (22). Darkness stimulates the excretion of melatonin while light suppresses its activity. Exposure to excessive light in the evening or too poor light during the day can disorder the body's normal melatonin cycles. For example, jet lag, shift work, and poor vision can disrupt melatonin cycles. In addition, some experts claim that exposure to low-frequency electromagnetic fields (common in household appliances) may disrupt normal cycles and production of melatonin.

Figure (22): Melatonin cycle

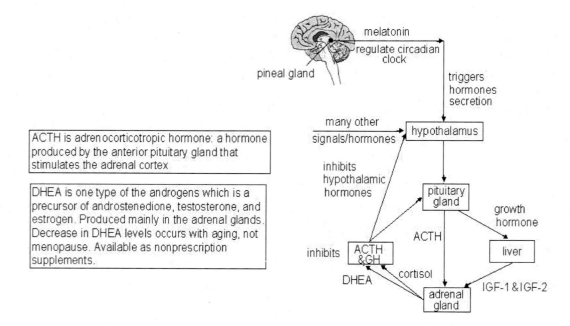

The chemical formula of melatonin shows that melanin is a good antioxidant to radicals such as OH, O2⁻, NO, Cl., -yl, nitroxide mediated polymerization, etc. For example, oxygen radicals can be bonded easily to H of the Ch3 molecule, and H in the HN molecule, Figure (23).

Figure (23): Chemical structure of melatonin

4-4-7-1 Roles of Melatonin in humans

- Heart Disease
  Low blood levels of melatonin are associated with heart disease. In addition, several animal studies suggest that melatonin may protect the heart from the damaging effects of ischemia (decreased blood flow and oxygen that often leads to a heart attack). However, researchers are unclear whether melatonin supplements may help prevent or treat heart disease in people.

- Antioxidant
  Melatonin is a direct scavenger of OH, O_2^-, and NO. Unlike other antioxidants, melatonin does not undergo redox cycling, the ability of a molecule to undergo reduction and oxidation repeatedly. Redox cycling may allow other antioxidants (such as vitamin C and Vitamin E) to regain their antioxidant properties.

- Cancer
  Because melatonin is antioxidant and does not undergo redox cycling, it is found that increased melatonin production has been proposed as a likely factor in the significantly lower cancer rates in night workers.

- Insomnia
  Since production of melatonin by the pineal gland is inhibited by light and permitted by darkness, studies suggest that melatonin supplements may help induce sleep in people with disrupted circadian rhythms (such as those suffering from jet lag or poor vision or those who work the night shift) and those with low melatonin levels (such as some elderly and individuals with schizophrenia). A review of clinical studies suggests that melatonin supplements may help prevent jet lag, particularly in people who cross five or more time zones.

- Osteoporosis

Melatonin has been shown to stimulate cells called osteoblasts that promote bone growth. Since melatonin levels may be lower in some older individuals such as postmenopausal women, current research is investigating whether decreased melatonin levels play a role in the development of osteoporosis, and whether treatment with melatonin can help treat this bone disease.

- Depression
  Clinical studies show that people who suffer from major depression or panic disorder have low levels of melatonin. Clinical studies have also found that melatonin may be useful in depression, especially associated with postmenopausal depression and anxiety. Melatonin use as anti-depressant medications can pose potential health risks and should only be used under direct supervision of a health provider.

- Rheumatoid Arthritis
  Melatonin levels are higher in a healthy individual than patients with rheumatoid arthritis. However, when arthritis patients were treated with the anti-inflammatory medications indomethacin, ibuprofen or naproxen, melatonin levels returned to normal. The chemical structure of melatonin resembles those medications, so researchers suspect that melatonin supplements may work similarly to these medications for people with rheumatoid arthritis. However, this should be thoroughly investigated.

- Alzheimer's disease
  There is limited study of melatonin for improving sleep disorders associated in patients with dementia and Alzheimer's disease (including nighttime agitation or poor sleep quality). It has been reported that natural melatonin levels are altered in people with Alzheimer's disease and dementia, although it remains unclear if supplementation with melatonin is beneficial. Further research is needed in this area before a firm judgment can be reached.

- Attention deficit hyperactivity disorder (ADHD)
  There is limited research of the use of melatonin in children with ADHD both on the treatment of ADHD and insomnia in ADHD children. A clear conclusion cannot be made at this time.

- Benzodiazepine tapering
  A small amount of research has examined the use of melatonin to assist with tapering or cessation of benzodiazepines such as diazepam (Valium) or lorazepam (Ativan). Although preliminary results are promising, further study is needed before a firm conclusion can be made.

- Glaucoma
It has been theorized that high doses of melatonin may increase intraocular pressure and the risk of glaucoma, age-related maculopathy (Age-Related Macular Degeneration is a degenerative maculopathy associated with progressive sight loss) and myopia (a condition where distant objects appear less clearly and those objects up close are seen clearly), or retinal damage. However, there is preliminary evidence that melatonin may actually decrease intraocular pressure in the eye, and it has been suggested as a possible therapy for glaucoma. Additional study is necessary in this area. Patients with glaucoma taking melatonin should be monitored by a healthcare professional.

3-5 Free radical theory of aging

In chemistry an unpaired electron is an electron that occupies an orbital on its own, rather than as part of an electron pair. Unpaired electrons or single electrons are uncommon in s- and p-block chemistry (see chapter 1), since the unpaired electron occupies a valence p orbital or a sp, sp^2 or sp^3 hybrid orbital. A lone pair is a valence electron pair without bonding or sharing with other atoms. They are found in the outermost electron shell of an atom, so lone pairs are a subset of a molecule's valence electrons.
There are many stable entities with unpaired electrons that do exist, e.g. the oxygen-molecule has two unpaired electrons and the nitric oxide molecule has one.

Denham Harman first proposed the free radical theory of aging in the 1950s, and in the 1970s extended the idea to implicate mitochondrial production of reactive oxygen species such as O_2^-, OH^- and H_2O_2. Aerobic organisms have elaborate defenses against reactive oxygen and radical species, which are produced during normal aerobic metabolism and in response to various stresses and diseases they have been linked with disorders ranging from neurodegenerative disease to cancer. The common radicals that are related to the free radical theory of aging are shown in table (2).

Table (2): Common radicals affecting theory of aging

Name	Chemical symbol	Source
Super oxide radical	O_2^-	Mitochondrion, enzymes, hydroquinoses
Hydrogen peroxide	H_2O_2	Enzymes, superoxide

Nitric oxide	NO⁻	Nitric oxide synthase enzymes, aerobic free radical chain reactions
Peroxynitrite	ONOO⁻	Superoxide + nitric oxide
Hydroxyl radical	OH⁻	Hydrogen peroxide, reduced transition metal ions
Hypochlorite	HOCl	myeloperoxidase
Transition metal ions	Fe^{++}, Cu^{++}, Mn^{++}, Ca^{++}, Fe^{+++}, etc.	Metalloprotein degradation, dietary overload
Singlet oxygen	$^1O_2$	Peroxynitrite, superoxide +hydrochlorite
Sodium and potassium radicals	Na^+, K^+	Drugs (for treatment of erythrocytes)

Oxygen is an example of a free radical that is normally stable. This means that oxygen radicals do not bond with each other, but oxygen can be quite reactive and unstable in the presence of other radicals. This is because oxygen has two unpaired electrons as shown below:

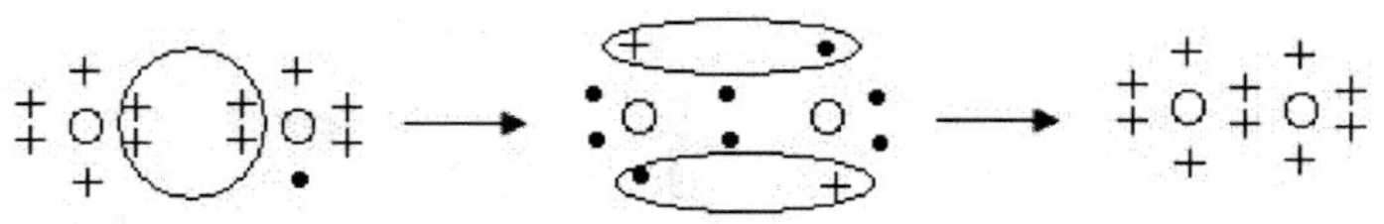

Radical oxygen (O_2^-) has the following configuration:

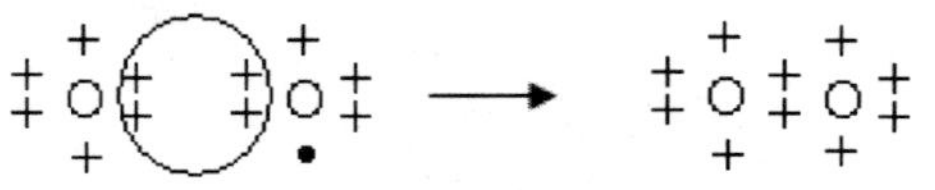

Radical oxygen will have the following reaction with other radical.

$$\overset{\cdot}{R} + O_2 \longrightarrow \overset{\cdot}{R} + \overset{\cdot}{O} + \overset{\cdot}{O} \longrightarrow R\,OO^{\cdot} \text{ (alkyldioxyle)}$$

$$ROO^{\cdot} + R\overset{\cdot}{H} \longrightarrow ROOH + \overset{\cdot}{R}_- \text{ (alkylhydroperoxide)}$$

$$\overset{\cdot}{R}_- + O_2 \longrightarrow R\_OO^{\cdot}$$

The subscript means a negative charge (anion).

In the above reaction, oxygen binds with organic free radicals ($R^{\cdot}$) to form a peroxyle radical ROO$^{\cdot}$, which in turn oxidizes an organic molecule such as RH$^{\cdot}$.

Table (3) shows some agents involved in the protection against oxidants and radicals.

Table (3): Agents for the protection against oxidants and radicxals

Name	Source	Function and benefit
Vitamin C	Vitamin C is found in many fresh vegetables and fruits, such as broccoli, green and red peppers, collard greens, brussel sprouts, cauliflower, lemon, cabbage, pineapples, strawberries, citrus fruits	An antioxidant vitamin needed for the formation of collagen to hold the cells together and for healthy teeth, gums and blood vessels; improves iron absorption and resistance to infection.
Vitamin E	The richest sources of vitamin E are cold-pressed crude vegetable oils, especially wheat germ, sunflower seeds, safflower, and soya bean oils. Eggs, butter, raw or sprouted seeds, and grains are moderately good sources	The main functions of vitamin E are to help protect the functioning of cells and the intracellular processes. It oxygenates the tissues and reduces the need for oxygen intake markedly. It prevents unsaturated fatty acids, sex hormones, and fat-soluble vitamins from being destroyed in the body by oxygen. It prevents the formation of excessive scar tissues.
Glutathione	Glutathione(GSH) is not found in foods, but certain foods can help your body to make more Glutathione: Avocado Asparagus, Broccoli Garlic, Raw Eggs, Spinach, tomatoes, Curcumin (Turmeric)and even Fresh unprocessed meats	Strengthen the Immune System, Act as the body's master antioxidant, Control the functions of other antioxidants, and works as the body's natural detoxifier. The lack of glutathione in turn accelerates the aging process as stated by James F. Balch MD and Phyllis A. Balch, http://www.springerlink.co/pp24331

Carotenoid	The best way to get dietary carotenoids is through natural sources such as orange and coloured fruits and vegetables. Some studies have shown that taking b-carotene. supplements may be harmful – particularly for smokers.	Carotenoids provide bright coloration, serve as antioxidants, and can be a source for vitamin a activity. Carotenoids are responsible for many of the red, orange, and yellow hues of plant leaves, fruits, and flowers, as well as the colors of some birds, insects, fish, and crustaceans. Some familiar examples of carotenoid coloration are the oranges of carrots and citrus fruits, the reds of peppers and tomatoes, and the pinks of flamingoes and salmon.
Bilirobin	Bilirubin is not found in food.It is the main bile pigment that is formed from the breakdown of heme in red blood cells. The broken down heme travels to the liver, where it is processed and secreted into the bile by the liver.	In adults or older children, bilirubin is measured to diagnose and/or monitor liver diseases, such as cirrhosis, hepatitis, or gallstones. Patients with sickle cell disease or other causes of hemolytic anemia may have episodes where excessive RBC destruction takes place, increasing bilirubin levels. Bilirubin is excellent antioxidant. Although bilirubin may be toxic to brain development in newborns, high bilirubin in older children and adults does not pose the same threat. In older children and adults, the "blood-brain barrier" is more developed and prevents bilirubin from crossing this barrier to the brain cells. Elevated bilirubin levels in children or adults, however, strongly suggest a medical condition that must be evaluated and treated.
Uric acid	Uric acid is the final oxidation (breakdown) product of purine metabolism and is excreted in urine. Purine is One of the groups of nitrogenous bases that are part of a nucleotide. Purines are adenine and gaunine, and are double-ring structures. See DNA structure, Chapter 1	Uric acid may be a marker of oxidative stress and may have a potential therapeutic role as an antioxidant. It is unclear whether elevated levels of uric acid in diseases associated with oxidative stress such as stroke and atherosclerosis are a protective response or a primary cause.
Ferritin	Ferritin is a protein that	The body has a "buffer" against

	stores iron and releases it in a controlled fashion	iron deficiency (if the blood has too little iron, ferritin can release more) and against iron surplus ((if the blood and tissues of the body have too much iron, ferritin can help to store the excess iron). Iron in blood hemoglobin is oxidized by oxygen which will be carried to cells and muscles.
Peroxiredoxins	Peroxiredoxins are a ubiquitous family of antioxidant enzymes that also control cytokine-induced peroxide levels and thereby mediate signal transduction in mammalian cells. They are biosynthesized in vivo.	Peroxiredoxins ((Prxs) are antioxidant proteins defenses which reduce H_2O_2 and peroxinitrite, and chaperones, which protect against protein unfolding and aggregation. Peroxiredoxins (Prxs) are a ubiquitous family of antioxidant enzymes that also control cytokine-induced peroxide levels which mediate signal transduction. Peroxiredoxins are a ubiquitous family of antioxidant enzymes that also control cytokine-induced peroxide levels which mediate signal transduction
Metallothioninine	They are synthesised primarily in the liver and kidneys.	They may provide protection against metal toxity which involved in regulation of physiological metals (Zn and Cu) and provide protection against oxidative stress.
Enzyme mimetic	They are produced in vivo and can be in vitro	Enzyme mimetic such as SOD/catalase mimetic can improve redox potential, and can prevent the oxidative-mediated damage associated with environmental stress.
PBN (phenyl N-tert-butylnitrone)	They are biosynthesized in vivo.	Phenyl-N-tert-butylnitrone (PBN) shows its major effect by scavenging free radicals formed in the ischemia and it has the ability to penetrate through the blood brain barrier easily

Here are some examples of antioxidants agents on oxidative species:

$$O_2^- + O_2^- + 2H^+ \xrightarrow{\text{SOD}} O_2 + H_2O_2$$

$$\downarrow$$

$$2H_2O_2 \text{ (peroxide)} \xrightarrow{\text{CAT}} O_2 + 2H_2O$$

$$\downarrow \text{ or}$$

$$H_2O_2 + 2GSH \longrightarrow 2H_2O + GSSG$$

First, the SOD change radicals O2- and H+ to stable oxygen and peroxides, and then catalase (CAT) changes the peroxide to oxygen and water. The peroxide can also with the effect of monomeric glutathione (GSH), be changed to water and glutathione disulfide (GSSG).

3-5-1 Rates of free radicals on aging

There is increasing evidence for the accumulation over time of damaged DNA and the modification of proteins and other molecules due to increasing radicals. It is calculated that endogenously generated oxygen free radicals make about 10,000 oxidative interactions with DNA per human cell per day (Ames et al, 1993). These alterations and damage to such vital molecules and cells would be expected to ultimately lead to deficiencies in normal functions in aging. The following protection and prevention schemes can reduce the number of oxidative interactions per day, and consequently delay the onset of age and age associated diseases:

- Antioxidant enzymes such superoxide dismutase and catalase and redox active molecules such as glutathione and thioredoxin.
- Vegetables and fruits in particular lutein* and lycopene** (very rich in vitamin beta-carotene), members of the carotenoids (yellow, orange and red pigments that occur widely in plants and animals often giving them a bright coloration) have received much attention as cancer preventive foodstuffs. Many studies have shown that a diet predominant in vegetable and fruits is associated with a reduced risk of many age-related serious diseases. There is a strong implication that fruits and vegetables are important in slowing the aging process.

- Restriction of caloric (sugar and starch) intake lowers steady-state levels of oxidative stress and damage, retards age-associated diseases and changes, and extends the maximum life-span in mammals. Many researchers believe the evidence to date shows the practice of caloric restriction will extend the healthy human life span, but consensus has not yet been reached on this topic. Table (4) shows foods of high rich calories.

*Lutein is in a green base of spinach, collard Greens, turnip, broccoli, celery, kale and radish.

** Lycopene is in a vegetable base of tomato, red pepper, D. salina, chlorella and spirulina

Table (4): Rich calories food

Food	Measurement	Carbohydrate in grams
Fruit cake	1 cake	783
Carrot cake	1 cake	775
Sheet cake with FRSTNG	1 cake	694
White cake	1 cake	670
Devil's cake	1 cake	645
Yellow cake	I cake	638
Pecan pie	1 pie	423
Cherry pie	1 pie	363
Apple pie	1 pie	330
Blueberry pie	1 pie	330
Cherry pie	1 pie	317
Ginger bread cake	1 cake	291
Honey	1 cup	279
Italian bread	1 loaf	256
French Vienna bread	1 loaf	230
Sugar brown	1 cup	212
Sugar white	1 cup	199
Danish pastry	1 ring	152
Figs dried	10 figs	122
Grape juice	6 Oz	96

3-6 Theories of aging

Senescence (old man in Latin) is the post-maturation decline in survivorship and fecundity that accompanies advancing age. Two main evolutionary theories have been evolved to account for senescence. They are:

- The mutation-accumulation theory. Deleterious mutations exerting their effects only late in life would tend to accumulate, because of their minimal effects on fitness. More precisely, exclusively late-acting deleterious mutations will attain higher equilibrium frequencies under mutation-selection balance than will mutations that act early, resulting in lower mean values for fitness components late in life. Medawar emphasized the possibility that this effect would be enhanced by selection of modifiers (such as better natural selection*) that postpone the age of onset of genetic diseases. The theory was proposed by Peter Medawar in 1952. His idea was that ageing was a matter of neglect. Nature is a highly competitive place and almost all animals in nature die before they attain old age. Therefore, there is not much motivation to keep the body fit for the long haul - not much selection pressure for traits that would maintain viability past the time when most animals would be dead anyway, killed by predators, disease or by accident.

* For example, the peppered moth exists in both light and dark colors in the United Kingdom, but during the industrial revolution many of the trees on which the moths rested became blackened by soot, giving the dark-colored moths an advantage in hiding from predators. This gave dark-colored moths a better chance of surviving to produce dark-colored offspring, and in just a few generations the majority of the moths were dark. This is the natural selection.

- Medawar's theory was further developed by George Williams in 1957, who noted that senescence may be causing many deaths, even if animals are not 'dying of old age.' In the earliest stages of senescence, an animal may lose a bit of its speed, and then predators will seize it first, while younger animals flee successfully. Or its immune system may decline, and it becomes the first to die of a new infection. Nature is such a competitive place, said Williams, (turning Medawar's argument back at him), that even a little bit of senescence can be fatal; hence natural selection does indeed care.The pleiotropy theory as suggested by Williams that many of the genes with beneficial effects on early fitness components have pleiotropy deleterious effects on late fitness components, but are nevertheless favoured by natural selection. Williams observed that mutations are advantageous early in life, but bear a cost at older ages. They will be positively selected for the antagonistic pleiotropy theory of aging. This theory is based on the decline with age in the effect of age-specific fitness-component changes on total fitness. Selection experiments in Drosophila and Tribolium support the pleiotropy theory, although one such experiment gave results that only bordered on significance, but the mutation-accumulation theory has never been tested.

- The most recent theory was suggested by Hamilton in 1966. Hamilton suggested that there would be a higher rate of incorporation of favorable mutation which increases survival at earlier ages within the productive period, compared with genes acting later. Over a long period of evolutionary time, this would have the effect of raising survival rates at early adult ages relative to survival rates later in life. An initially non-senescent life-history, with constant adult C values, would then become converted to one which exhibits senescence, Figure (24).

Figure (24): Progress of senescence

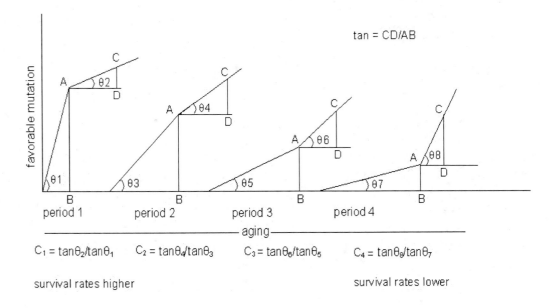

- Disposable soma theory is unlike the previous three theories: It does not link genes as being involved in the aging process but focuses on the idea that cell maintenance (DNA repair, protein turnover, antioxidant defenses, etc) is costly, given the fact that extrinsic mortality is extremely high in the wild. Extrinsic mortality is different than intrinsic mortality. Extrinsic mortality is imposed by the environment and is outside the control of the organisms (predation or weather). In the intrinsic theory the organism can exert some control over the short run which is subject to selective control over longer period. In most models of growth and development, mortality is treated as extrinsic and therefore not subject to selection (Charnove & Berrigan, 1993). For example, nine out of ten newborn wild mice will die before the age of 10 months* – it would make little sense to use precious metabolic resources to maintain the soma beyond the expected lifetime of the organism. Using extra energy to increase reproductive capacity will be more beneficial from an evolutionary standpoint because it will enhance the fitness of that individual. This means that the higher the extrinsic mortality rate for a species, the less energy should be invested in somatic maintenance. For example, animals with adaptations that offer increased protection from extrinsic mortality have longer life spans. Some examples are birds and bats that have wings allowing them to escape dangerous situations and turtles that have sturdy protective shells. It was found that opossums living on an island where extrinsic mortality was low, aged at a slower rate than opossums living on the mainland where there was significant risk of death by predation. There was no evidence that other environmental factors (i.e. parasitism, disease, decreased food availability) contributed to this difference.

* Austad, S.N. 1997. Comparative aging and life histories in mammals. Exp Gerontol 32:23-38.

3-7 NAD$^+$ and NADH in aging

Numerous studies have suggested that NAD$^+$ and NADH mediate multiple major biological processes, including calcium homeostasis, energy metabolism, mitochondrial functions, cell death and aging. Nicotinamide adenine dinucleotide (NAD$^+$), is a coenzyme found in all living cells. The compound is a dinucleotide, since it consists of two nucleotides joined through their phosphate groups, with one nucleotide containing an adenine base and the other containing nicotinamide, Figure (25) repeated for convenience.

Figure (25): Chemical formula of NAD$^+$

Nicotinamide Adenine Dinucleotide (NAD)$^+$

In metabolism the compound accepts or donates electrons in redox reactions. In metabolism, since NAD$^+$ is a cation, it is involved in redox reactions, taking electrons from one reaction to another. The coenzyme is therefore found in two forms in cells: NAD$^+$ is an oxidizing agent and NADH is a reducing agent, Figure (26).

Figure (26): Reduction and oxidation between NAD$^+$ and NADH

$$NAD^+ + H^+ + 2e \qquad\qquad NADH$$

NAD+ and NADH have emerged as novel, fundamental regulators of electrolytes particularly calcium homeostasis. It appears that most of the components in the

260

metabolic pathways of NAD+ and NADH, including poly (ADP-ribose), ADP-ribose, cyclic ADP-ribose, O-acetyl-ADP-ribose, nicotinamide and kynurenine, can produce significant biological effects. Numerous studies have suggested that NAD+ and NADH mediate multiple major biological processes, including calcium homeostasis, energy metabolism, mitochondrial functions, cell death and aging. Studies also proposed that NAD+ and NADH are fundamental mediators of brain functions, brain senescence and multiple brain diseases.

NAD+ and NADH play an important role in removing acetyl groups which are commonly transferred to conenzyme A (CoA) or from acetyle-CoA. Acetyl-CoA is an intermediate both in the biological synthesis and in the breakdown of many organic molecules. CoA is the final stage of the process of glycolysis in the cytoplasm where two stages of glycolysis take place: energy investing with NAD+ is one product, and energy harvesting with ATP is another product, Figure (27).

Figure (27): Glycolysis of glucose producing CoA

glycolysis in the cytoplasm

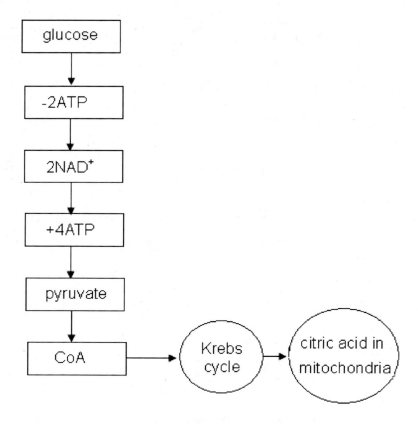

At the output of the glycolysis process, there is also another product, the coenzyme CoA. The role of NAD+ and NADH, in conjunction with SIR2 (Sirtuins are hypothesized to play a key role in an organism's response to stresses, such as heat or starvation, and to be responsible for the lifespan-extending effects of calorie restrictions) are to remove CoA at the end of the glycolysis process. SIR stands for Silent Information Regulator genes. The various sirtuins in mammals are named SIRT1-SIRT7. SIRT1 is the mammalian homolog (sometimes claimed to be ortholog) of SIR2. The discovery that SIR2 requires NAD for its activity immediately suggested a link between SIR2 activity and caloric restriction. This link was strengthened by the observation that life span extension by caloric restriction requires SIR2 protein. Caloric restriction is likely to reduce the carbon flow through glycolysis and result in more free cytoplasmic NAD. SIR2 could act as a sensor of NAD levels within the nucleus and reduce it back to normal. If NAD is higher than normal, the rate of aging is decreased, Figure (28).

Figure (28): SIR2 controlling NAD

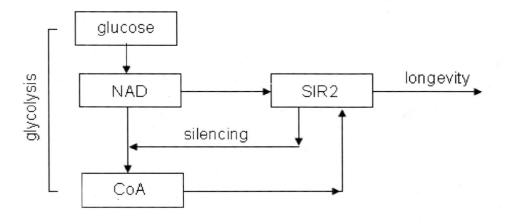

3-8 Telomere and telomerase and life span

The Nobel Prize in Physiology or Medicine is awarded this year (2009) to three scientists who have solved a major problem in biology: how the chromosomes can be copied in a complete way during cell divisions and how they are protected against degradation. The Nobel Laureates have shown that the solution is to be found in the ends of the chromosomes – the telomeres – and in an enzyme that forms them – telomerase. Telomeres are in the spotlight of modern biology. Whether the subject at hand is cancer, gene regulation, organismal aging, or the cloning of mammals, much seems to depend on what happens at the ends of chromosomes.

262

Before we discuss the subject in detail, we should know the structure of chromosomes, DNA, and the cell, Figure (29).

Figure (29): Structure of chromosomes, DNA and cell

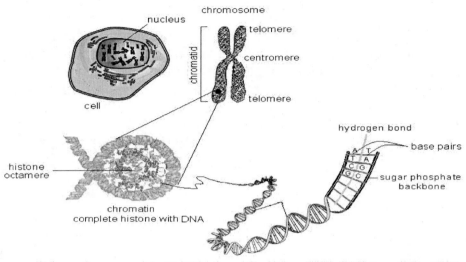

histone octamer consists of H3-H4 dimer (H3 + H4) andH2A - H2B dimer (H2A + H2A)

Chromosomes end with two telomeres, Figure (29). If the telomeres are shortened, cells and human age. Conversely, if telomerase activity is high, or if the telomeres are extended, chromosome (and DNA) are maintained, and cellular senescence is delayed. In contrast, if telomeres are damaged or defected, cancer and certain inherited disease may be developed. Scientists found that when a cell is about to divide, the DNA molecules, which contain the four bases that form the genetic code, are copied, base by base, by DNA polymerase enzymes. However, for one of the two DNA strands, a problem exists in that the very end of the strand cannot be copied, therefore, the chromosomes should be shortened every time a cell divides, Figure (30).

Figure (30): Shortened DNA and thus chromosome due to damaged or shortened telomere

263

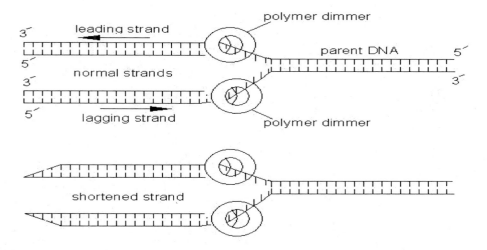

Luckily, these problems were solved when this year's Nobel Laureates (Elizabeth Blackburn, Jack Szostak, and Carol Greider) discovered how the telomere functions and found the enzyme that copies it.

3-9 Aging of cells

During his experiment, Jack Szostak observed that a linear DNA molecule, a type of minichromosome, is rapidly degraded when introduced into yeast cells. In fact, all DNA molecules are degraded by time after replication, but by putting the DNA molecules in yeast cells, they delay faster. The question is how to delay or prevent the degradation. If this is achieved, the span of life will be elongated and the aging process will be delayed.

3-9-1 Reversal of aging

Elizabeth Blackburn mapped DNA sequences. When studying the chromosomes of *Tetrahymena*, a unicellular ciliate organism. She identified a DNA sequence that was repeated several times at the ends of the chromosomes. The function of this sequence, CCCCAA (C for cytosine and A for adenine, both are nucleotides of the DNA), was unclear. Blackburn and Szostak decided to perform an experiment that would implement their discoveries jointly. From the DNA of *Tetrahymena*, Blackburn isolated the CCCCAA sequence. Szostak coupled it to the minichromosomes and put them back into yeast cells. Reslts were amazing - the telomere DNA sequence protected the minichromosomes from degradation, Figure (31). This was the start of the reversal of aging.

Figure (31): Szostack and Blackburn joint experiment

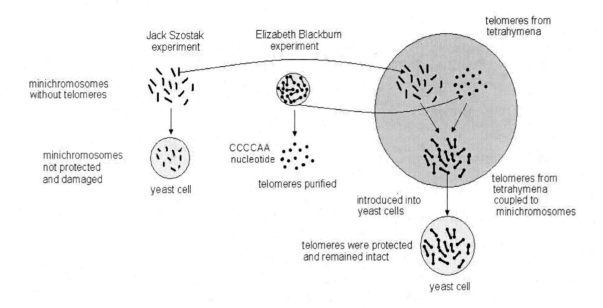

minichromosomes
without telomeres

Jack Szostak
experiment

Elizabeth Blackburn
experiment

telomeres from
tetrahymena

minichromosomes
not protected
and damaged

yeast cell

CCCCAA
nucleotide

telomeres purified

introduced into
yeast cells

telomeres from
tetrahymena
coupled to
minichromosomes

telomeres were protected
and remained intact

yeast cell

Blackburn and her graduate student Carol Greider started to investigate if the formation of telomere DNA could be due to an unknown enzyme. Greider discovered signs of enzymatic activity in a cell extract. Greider and Blackburn named the enzyme telomerase. The telomerase consists of RNA nucleotides CCCCAA and protein. Telomerase extends telomere DNA, providing a platform that enables DNA polymerases to copy the entire length of the chromosome without missing the very end portion, Figure (32).

Figure (32): Effect of telomerase on cell division

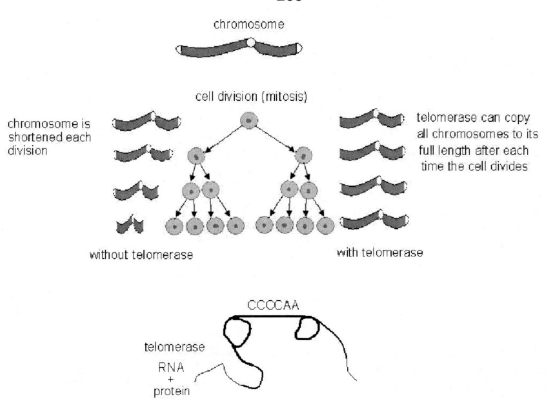

With gradual shortening of the telomeres and chromosomes cells grew poorly and eventually stopped dividing. This could lead to premature cellular ageing – senescence. Scientists concluded that functional telomeres could prevent chromosomal damage and delay cellular senescence. They showed that the senescence of human cells is also delayed by telomerase.

3-10 Senescence and titrating nuclear factors

A number of genetic components of aging have been identified using different organisms, ranging from the simple budding yeast Saccharomyces cerevisiae to worms such as Caenorhabditis elegans and fruit flies (Drosophila melanogaster). Study of these organisms has revealed the presence of at least two conserved aging pathways. One of these pathways involves the gene SIR2 (Silent Information Regulator), a NAD^+ and protein (histone). In yeast, SIR2 is required for genomic silencing at three loci: the yeast mating loci, the telomeres and the ribosomal DNA (rDNA), Figure (33).

Figure (33): Function of SIR2 and the three loci

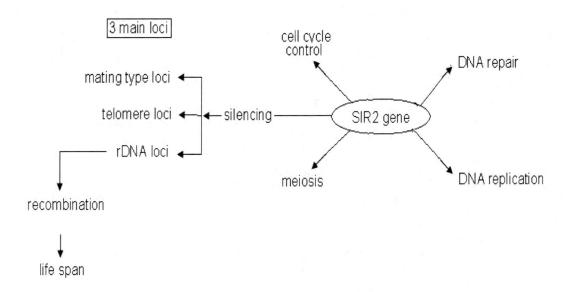

SIR2 seems to affect life span by silencing the accumulation of the extrachromosomal rDNA circles (ERC), produced from the combination event of nuclear rDNA copies. ERCs induce replication of senescence of rDNA in the mother cell. SIR2 mediated silencing of rDNA copies inhibits recombination and therefore ERC formation. The ribosomal DNA (rDNA) or recombinant DNA (recombinant DNA is the addition of relevant DNA or a bacteria plasmid into an existing organismal DNA), is transcripted into sequences separated by nontranscribed spaces, Figure (34).

Figure (34): Transcription of rDNA

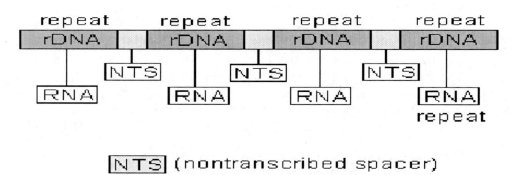

SIR2 silences the rDNA during the repetition of rDNA as shown in Figure (35).

Figure (35): Segregation of rDNA by SIR2

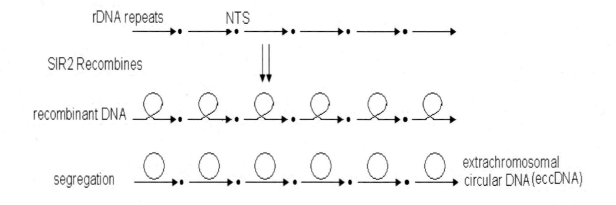

titration = essential cellular factors death

Some species of yeast replicative aging may be partially caused by homologous recobination between rDNA repeats: excision of rDNA repeats results in the formation of extrachromosomal rDNA circles (eccDNAs). These eccDNAs replicate and preferentially segregate to the mother cell during cell division, and are believed to result in cellular senescence by titrating away (competing for) essential nuclear factors.

SIR2 deacetylases the histone protein through NAD^+ (a cofactor) to produce nicotinamide and O-acetyle-ADP-ribose as the following equation:

$$NAD^+ + \text{acetylated histone} \xrightarrow{SIR2} \text{nicotinamide} + O\text{-acetyle-ADP-ribose} + \text{histone}$$

Figure (36) shows acetylated and deacetylated histone

Figure (36): Acetaylation and deacetylation of histone (Lysine)

deacetylated histone   acetylated histone

SIR2 deacetylates protein with NAD+ to produce nicotinamide and O-acetyle-ADP-ribose is shown in Figure (37).

Figure (37): Deacetylation of protein by SIR2

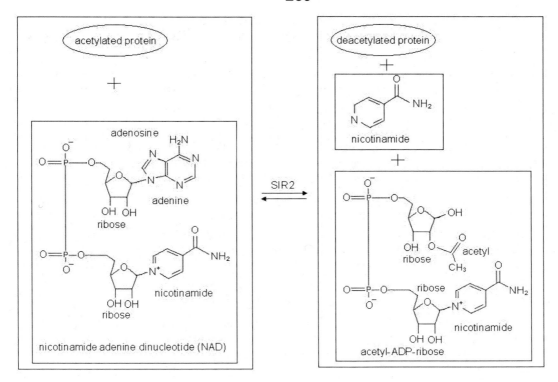

Acetylation of protein has a short life span, whereas deacetylation has a great extended life span. In other words, SIR2 reduction or deletion shortens the life span, and additional or extra SIR2 will elongate the life span.

3-11 Ras protein

The cell cycle consists of a G1 growth phase, a S phase where the DNA of the cell is replicated (doubled), a G2 second growth phase, and finally an M (mitosis) phase where the cell divides forming a new cell. Certain regulatory molecules bind to cell membrane receptors that have an intercellular domain that is called a tyrosine kinase. Tyrosine is one of the 20 amino acids that are used by cells to manufacture proteins and kinase means phosphotransferase. A tyrosine kinase is an enzyme that can transfer a phosphate group from ATP to a tyrosine residue in a protein as the following equation:

ATP + a protein tyrosine = ADP + protein tyrosine phosphate

Many enzymes and receptors are switched "on" or "off" by phosphorylation and dephosphorylation. Scientists worked out most of the details of how ras proteins control the operation of the cell cycle. It was subsequently found that single point mutation of the ras gene can lead to the formation of an oncogene and a cancer causing ras protein.
Ras protein ultimately leads to the phosphorylation of transcription factors in the nucleus, which in turns alter gene expression.

The binding of a growth factor molecule to a receptor on the cell membrane is the stimulus that begins the process. The ras protein is normally in the inactive state when it is associated with a molecule called GDP. Upon being activated by the growth factor stimulus, it switches GDP for the more active GTP (GDP and GTP stand for guanosine diphosphate and guanosine triphosphate nucleotide). Before the ras protein can become functional, it must receive a small molecule with the aid of an enzyme called farnesyl transferase (farnesyl transferase is responsible for activating RAS). At this point, the ras protein attaches to the cell membrane and initiates a cascade of enzymatic reactions. Eventually a protein enters the nucleus, where it activates a "transcription factor", which in turn activates the cyclin D gene. The cyclin D protein that is formed stimulates the progression of the cell cycle into the S phase and through the remainder of the cycle. This cyclin forms a complex with and functions as a regulatory subunit of the cyclins group whose activity is required for cell cycle G1/S transition. In normal cells, the ras protein reverts to the inactive state after transmitting the signal by replacing GTP to GDP, Figure (38). However, a mutation in the ras gene results in the formation of a ras protein that no longer has the enzyme to convert GTP to GDP. Therefore, the ras protein remains in the active state even with no stimulus from growth factors. The ras protein then sends continuous signals to keep the cell cycle running with no checks and balances. The result is excessive cell proliferation and cancer.

Figure (38): Ras protein and gene expression

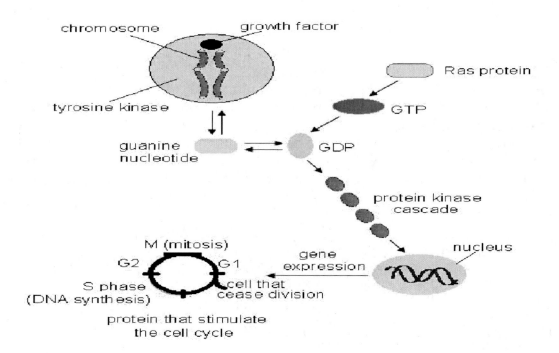

3-11-1 Ras protein and aging

Ras1, together with Ras2, forms a subgroup of Ras proteins regulating the progression of Saccharomyces Cervisiae cells through the G1 phase of the cell cycle. Ras1 protein is more expressed than Ras2 during growth on aerobic glucose compared to non fermented glucose. Both proteins appear to be components of a division-making mechanism linking nutrients availability to progression of yeast cells from early G1 to subsequent steps of the cell cycle.

Over-expression of RAS2 led to a 30% increase in the life-span on average and postponed the senescence-related increase in generation time seen during Saccharomyces Cervisiae yeast aging. No life-span extension was obtained by overexpression of RAS1. However, deletion of RAS1 prolonged the life-span. These results suggest that RAS1 and RAS2 play reciprocal (reversal) roles in determining yeast longevity. The major known pathway through which Ras proteins function in yeast involves stimulation of adenylate cyclase. Adenylate cyclase converts GTP to cAMP (cyclic adenosine mono phosphate), Figure (39). A balanced expression between Ras1 and Ras2 would not extend the life-span. Therefore, expression of Ras2 should be increased so that life-span can be extended in Saccharomyces Cervisiae yeast.

Figure (39): Conversion of GDP to cAMP by adenylate cyclase

3-12 DNA damage theory of aging

Animal cell nucleus has a complex substance known as chromatin — which consists of DNA, five histone proteins and some non-histone proteins. Histone protein is not coagulated by heat and is composed of a high proportion of the basic amino acids lysine and arginine, which are positively-charged at physiological pH. Because DNA is negatively-charged (due to phosphate groups), the positively-charged histones readily bind to DNA, Figure (40).

Figure (40): Binding of positively charged arginine and negatively charged DNA

Four of the histones (H2A, H2B, H3 and H4), Figure (41), compact DNA about six-fold into bead-like nucleosomes. A fifth histone (H1) binds to the DNA between nucleosomes, causing a second-order compacting of the DNA. With age, compacting of chromatin increases which leads to a decline in gene expression. A 50% reduction in chromatin-associated RNA polymerase II activity has been demonstrated in the brains of old rats, see http://www.benbest.com/lifeext/aging.html#dna.

Figure (41): Compacting of DNA by histones

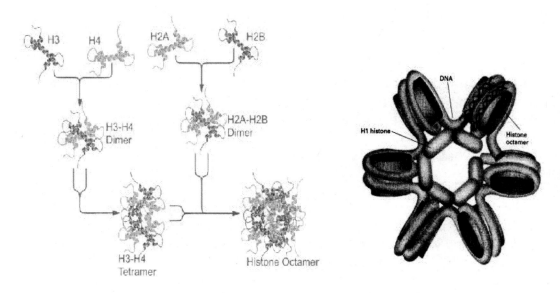

DNA damage includes altered bases, mismatched base pairs, strand cross-linking, and both single- and double-strand breaks. The three main repair pathways are nucleotide excision repair, base excision repair, and direct reversal of damage. As the molecular mechanisms of DNA repair pathways became clearer during the 1980s and 1990s, it became necessary to study this question in details, but the mechanism of DNA damage and repair in aging remains unclear.

Early work on DNA repair focused mainly on the repair of pyrimidine dimers (dimers means binding of two identical molecules) in DNA (note that pyrimidine are thymines and cytosines, and purines are adenines and guanines) dimers are produced by the cross-linking of two adjacent pyrimidine bases when DNA is exposed to ultraviolet light. It became clear that such dimers may not be the most abundant DNA damage in vivo, as few cells are actually exposed to ultraviolet light. It is now recognized that altered bases and dimers cross linking due to oxidative stress occur much more frequently than the damage due to ultraviolet exposure. It has been estimated that as many as 100,000 oxidized bases may be generated in DNA per cell, per day. Such damage is repaired by a pathway known as base excision repair, which begins by removal of the damaged base, followed by DNA breakage at the site of the missing base, removal of the remaining damage, and replacement of the missing nucleotide(s) (phosphate plus base plus ribose) by DNA polymerase β. DNA polymerase beta performs base excision repair required for DNA maintenance, replication, recombination, and drug resistance. This enzyme is absolutely essential for mammalian viability, although cells lacking it can be grown in culture. This suggests that a certain amount of DNA damage can be tolerated in single cells grown in culture, but that the combined effect of many damaged cells in one or more critical tissues is not tolerable.

The general results from recent experiments include: (1) mutation frequency does increase with age, (2) mutations tend to be greater in proliferating tissues (benign or malignant tumors) than in nonproliferating tissues,and (3) the nature of the mutations varies with age and tissue examined. All mutations and chromosomal rearrangements accumulate with age in most examined tissues, and it is assumed that the latter has more serious implications for the aging individual.

Another important biomarker is cellular morphology (change in cell shape due to mutation). The progressive morphological changes cells endure while they age in vitro were particularly well-studied by Klaus Bayreuther and his colleagues (e.g., Bayreuther et al., 1988)). In brief, senescent cells are bigger and a senescent population has more diverse morphotypes than cells at earlier cumulative population doubling (we shall talk about it later). In fact, a confluent senescent culture has a smaller cellular density than a confluent young culture, though this also occurs because senescent cells are more sensitive to cell-cell contact inhibition.

Most common mutations in nucleus DNA and mitochondrion DNA is shown in Figure (42).

Figure (42): different types of mutation

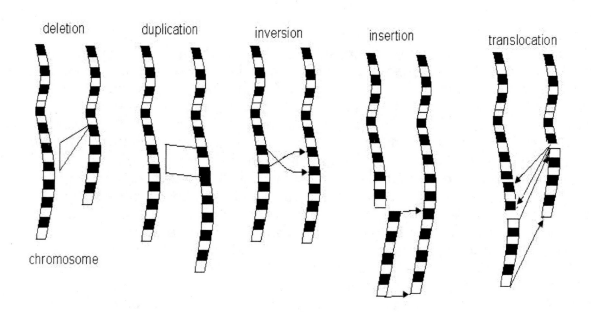

Mutations in mitochondrial DNA also play an important role in senescence. Before we talk about the mutation in mitochondrial DNA, let's talk about mitochondria.

Mitochondria are the cellular organelles that generate energy from aerobic metabolism. Glucose and other food molecules are oxidized to carbon dioxide and water. The energy released is stored in the form of adenosine triphosphate (ATP).

Mitochondria are enclosed by a double membrane. The intermembrane space between the two membranes is separated from the cytosol outside and the mitochondrial matrix inside. Pyruvate that is produced by glycolysis in the cytosol is transported into the matrix where it enters into the Krebs cycle to form nicotinamide adenine dinucleotide (NADH) and flavin adenine dinucleotide (FADH2). Several protein complexes of the inner membrane form an electron transfer chain where electrons from NADH and FADH2 are transferred to oxygen. This releases energy that is used by the protein complexes to transport protons into the intermembrane space. The ATP synthetase complexes use the flow back of these protons into the mitochondria to produce ATP.

Oxygen free radicals produced by the respiratory chain in mitochondria can damage mtDNA and thus increase the mutation rate.

Mutations in mitochondrial DNA also increase with age, particularly in the form of deletions (deletions of large segments of the genome of mitochondrion) . The in vivo cellular impact of age-associated mitochondrial DNA mutations is unknown. It is hypothesized that mitochondrial DNA deletion mutations contribute to the fiber atrophy and loss that cause sarcopenia and the age-related decline of muscle mass and function. The human mitochondrial genome is extremely small compared with the nuclear genome, and mitochondrial genetics presents unique clinical and experimental challenges. Defects in the mitochondrial genome accumulate with age and are thought to have a causal role in aging processes. The mitochondrial genome is a closed circular DNA molecule and it encodes 22 tRNAs, two rRNAs and 13 polypeptides of the electron transport system. Two to ten copies of the genome are found in each mitochondrion. The mitochondrial genome is susceptible to mutation damage because the circular DNA is located near the source of reactive oxygen species production, it lacks histone protection and DNA repair systems are limited in mitochondria. These mutations include duplications of regions of the mitochondrial genome, point mutations and deletions of large segments of the genome. mtDNA deletion mutations accumulate with age and are commonly detected in post-mitotic tissues, such as brain, heart and skeletal muscle, which rely heavily on oxidative metabolism. Despite the diminutive size of the mitochondrial genome, mitochondrial DNA (mtDNA) mutations are an important cause of inherited disease. Recent years have witnessed considerable progress in understanding basic mitochondrial genetics and the relationship between inherited mutations and disease phenotypes, as well as identifying acquired mtDNA mutations in both ageing and cancer.

Diseases resulting from mutations in mitochondrial genes are usually maternally inherited. Ova contain mitochondria, but spermatozoa contain few, hence, the mitochondrial content of zygotes is derived almost entirely from the ovum. Thus mitochondrial DNA is transmitted entirely by females. Affected females transmit the disease to all their offspring, (male and female), however, daughters and not sons pass the disease further to their progeny, see http://www.pathopedia-india.com/MutationsMitochondrialGenes.htm.

3-13 WRN geneene and Werner's syndrome

The WRN gene provides instructions for producing the Werner protein, which plays an important role in repairing damaged DNA. The Werner protein performs as a type of enzyme called a helicase. Helicase enzymes generally unwind and separate a damaged double-stranded DNA. The Werner protein also functions as an enzyme called an exonuclease. Exonucleases cut the broken ends of damaged DNA by removing DNA building blocks (nucleotides). The study suggests that the Werner protein first unwinds the DNA and then removes damaged DNA structures that have been accidentally produced.

Werner syndrome is a premature aging disease that begins in teenagers or early adulthood and results in the appearance of old age by 30-40 years of age. Its physical characteristics may include short stature and other features usually developing during adulthood such as graying and loss of hair; a hoarse voice; and thin, hardened skin. They may also have a facial appearance described as "bird-like", wrinkled skin, baldness, cataracts (cloudy lenses), muscular atrophy (decrease in the mass of the muscle) and a tendency to diabetes mellitus, among others.

The disorder is inherited and transmitted as an autosomal recessive trait (An autosome chromosome is not a sex hormone. In humans, there are 22 pairs of autosomes, but there is only one sex chromosome). Cells from WS patients have a shorter lifespan in culture than normal cells. This autosome chromosome is a predicted helicase belonging to the RecQ family (three types of RecQ helicases possess a novel strand pairing activity). Werner syndrome is also known as progeria adultorum, progeria of the adult, and pangeria. Werner syndrome is the most common of the premature aging disorders. Progeria can also refer to Hutchinson-Gilford syndrome, which is described as a lamin A gene defect and has onset early in life. The term progeria is derived from a Greek term meaning "before old age." Figure (43) shows pictures of both Werner syndrome and progeria syndrome.

Figure 43): Persons with Werner syndrome and progeria syndrome

Werner syndrome

Progeria syndrome

same person

age 15 years   age 48 years

http://www.bioportfolio.com/search/werner
_syndrome_photos.html

http://www.carolguze.com/images/clinical
/progeria.jpg

Some studies showed that Werner syndrome is connected with excessive synthesis of collagen types I and III. Collagen has great tensile strength, and is the main component of fascia, cartilage, ligaments, tendons, bone and skin. Along with soft keratin, it is responsible for skin strength and elasticity, and its degradation leads to wrinkles that accompany aging. The gelatin used in food and industry is derived from the partial hydrolysis of collagen. Collagen types I and II are dependent on elevated messenger RNA (mRNA) levels. The collagenase level is also increased several times with Werner syndrome.

3-13-1 Symptoms of Werner syndrome

- Short stature, usually between 1.40 and1.60 m.
- Muscle atrophy is noted.
- Thin skin is present on the body surfaces. Wrinkling and aging of the face occurs.
- The skull is relatively large, with a disproportionate size of lower part of the face.
- A sclerodermalike (thickening and hardening of skin) appearance with nose and lip atrophy is typical.
- Loss of subcutaneous fat associated with ulceration, noticed on the shins and feet.
- The nose is pinched, and the cheeks are sunken because of fat loss, which look likes the "birdlike" facial appearance.
- Knees are bent down because of fat loss and gravity.
- Calluses (calluses and corns are a discrete form of hyperkeratosis), hyperkeratosis, and ulcerations on the soles are present mainly over bony prominences.
- Graying of the hair and loss of hair usually are observed.

- Muscular dystrophy, Duchenne muscular dystrophy which usually appears in male children before age 6 and may be visible in early infancy. This is progressive proximal muscle weakness of the legs and pelvis, reflex sympathetic dystrophy type II which could lead to nerve damage, retinal dystrophy, and myotonic dystrophy (difficulty realaxing a muscle that could build cataracts,heart conduction defects, endocrine changes).
- A high-pitched voice is characteristic.
- Diabetes mellitus is characteristic.
- Neoplasia which could generate many types of cancers such as sarcomas, carcinomas of the thyroid, hematologic malignancies, cutaneous malignancies, including malignant melanoma, meningiomas, and squamous cell carcinomas. You may refer to the book of Chemistry, Biology and Cancer: The Bond by the author of this book.

Other symptoms which are not observable are:

- Hyperthyreosis
- Vascular changes (arteriosclerotic type) which could lead to cardiovascular disease and stroke, the main causes of mortality in industrialized countries.
- Pituitary dysfunction which could lead to hypopiyuitarism. Hypopituitarism is the decreased secretion of one or more of the eight hormones produced by the pituitary gland.
- Soft tissue calcification (do not confuse soft tissue calcification with bone calcification. Bone calcification deals with bone tumor)
- Hypogonadism or agonadism and premature menopause.

3-14 Late-Life mortality deceleration

Gompertz law (1825) for advanced ages was modified b British researchers Greenwood and Irwin in 1939. Gompertz expressed the relationship between the age specific mortality rate, m_x and age itself x, for ages over 15 in the formula:

$M_x = BC^x$ for x > 15 1

Where B and C are constant.

Equation 1 can be transferred to:

Log (m_x) = log B + x log C 2

Now equation 2 is a linear relationship between log (m_x) and x in which the slop of the line is log C, Figure (44).

Figure (44): Gompertz linear representation between mortality rate and age

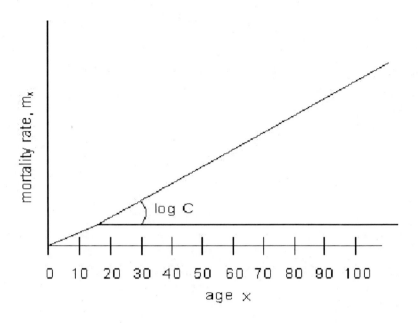

Greenwood and Irwin stated:

- "The increase of mortality rate with age advances at a slackening rate, that nearly all, perhaps all, methods of graduation of the type of Gompertz's formula overstate senile mortality".
- "... Possibility that with advancing age the rate of mortality asymptotes to finite value.
- "... The limiting values of q_x are 0.439 for women and 0.544 for men". The values represent one-year probability of death, q_x, at extreme ages.

The difference between Greenwood and Irwin, and Gompertz in mortality deceleration is shown in Figure (45).

Figure (45): Mortality deceleration by Greenwood and Irwin, and Gompertz

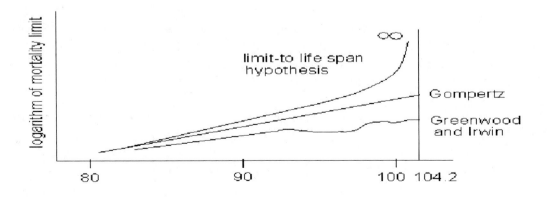

Mortality deceleration develops in all species. However, no mortality deceleration was reported for rodent (Austad, 2001) and baboons (Bronkowski et al, 2002).

3-14-1 Reasons of mortality deceleration

- Low risks of death for older people due to less risky behaviour
- Exhaustion of organism's redundancy
- Population heterogeneity (subpopulation with the higher injury level due to vigour in life, die more rapidly)
- Evolutionary explanation

To conclude:

- All suggestions by Greenwood, Irvin and Gompertz regarding the mortality deceleration appear to be not that strong - cohort mortality at advanced ages grows exponentially up to age 105 years.
- Mortality deceleration at advanced ages should be less expressed for data of higher quality. Records for persons applied to SSN (Social Security Number) in the Southern States, Hawaii and Peurto Rico were suggested to have low quality.
- Life expectancy at age 80 depends on the month of birth. Assuming that debilitation in utero or in the first year of life increases the infant's susceptibility to diseases as an adult, based on that fact, people born between October and December (autumn) have higher birth weights than those born in other seasons.

3-15 Reliability theory and redundancy exhaustion

The Reliability theory of aging and longevity is a scientific approach aimed to gain theoretical insights into mechanisms of biological aging and species survival patterns by applying a general theory of systems failure, known as the reliability theory.

Reliability is the probability that a human body will perform its intended function satisfactorily for its intended life under specified environmental and biological conditions such as natural selection and genetic inheritance.

Reliability theory allows researchers to predict the age-related failure kinetics for a system of given architecture (reliability structure) and the given reliability of its components. Applications of reliability-theory approach to the problem of biological aging and species longevity lead to the following conclusions:

Redundancy is a key for understanding aging and the systematic nature of aging in particular. Systems, which are redundant in numbers of irreplaceable elements, do deteriorate (that is, age) over time, even if they are built of non-aging elements. Redundancy has two important consequences:

- At very high ages, the phenomenon of aging apparently disappears and the failure rate disappears, i.e., failure rate levels off, Gavrilov and Gavrilova 2005.
- The system with different initial levels of redundancy will have very different failure rates in early life, but these differences will eventually vanish as failure rates approach the upper limit determined by the rate of elements loss. This is because the compensation law of mortality (late-life mortality convergence), (Gavrilov and Gavrilova, 1979), states that the relative differences in death rates between different populations of the same biological species decrease with age. This is because the higher initial death rates in disadvantaged populations are compensated by lower pace of mortality increase with age, Figure (46).

Figure (46): Compensation law of mortality. Convergence of mortality rate in different population at advanced ages

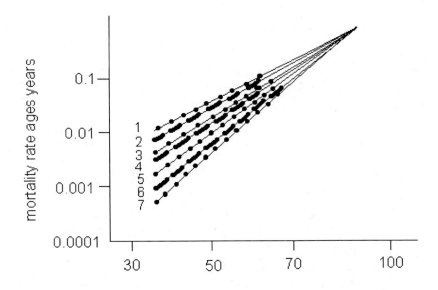

Gavrilove and Gavrilova stated that aging rates are affected by numbers of redundancies. For example, if redundancy is zero, then there will be no aging, i.e., the higher the number of redundancies is the higher the aging rate, Figure (47).

Figure (47): Aging rates and redundancies

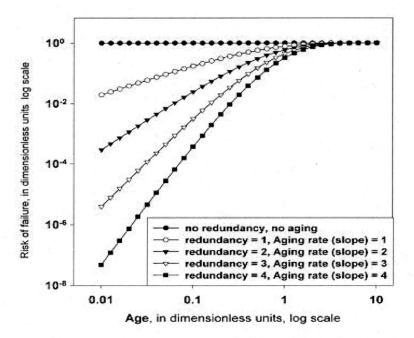

Gavrilove and Gavrilova pointed out that an apparent aging rate or expression of aging (measured as age difference in failure rates, including death rates) is higher for systems with higher redundancy levels.

3-16 Fisher's reproductive value

Because mortality accumulates with age, Fisher proposed that the strength of selection (mainly natural selection) acting on survival should increase from birth up to the age of first reproduction. Hamilton later theorized that the strength of selection acting on survival should not change from birth to age at first reproduction. As organisms in nature do not live in uniform environments but, rather, experience periodic stress, It was hypothesized that resistance to environmental stress should increase (Fisher) or remain constant (Hamilton) from birth to age at first reproduction. Fisher's fundamental theorem of natural selection, that the rate of change of fitness is given by the additive genetic variance of fitness, has generated much discussion since its appearance in 1930. Fisher tried to capture in the formula the change in population fitness attributable to changes of allele (genomes) frequencies, when all else is not included.

Before we talk about the mathematical model of Fisher's reproductive vale lets discuss the life table of humans, Table (5).

Table (5): Life table of the residents of the 50 States and the District of Columbia

Exact age	Male			Female		
	Death probability [a]	Number of lives [b]	Life expectancy	Death probability [a]	Number of lives [b]	Life expectancy
0	0.007566	100,000	74.81	0.006156	100,000	79.95
1	0.000522	99,243	74.38	0.000416	99,384	79.45
10	0.000096	99,033	65.53	0.000107	99,217	70.58
20	0.001314	98,474	55.87	0.000454	98,961	60.74
30	0.001413	97,082	46.60	0.000636	98,453	51.03
40	0.002436	95,450	37.30	0.001442	97,556	41.45
50	0.005730	91,985	28.49	0.003295	95,432	32.24
60	0.011908	84,891	20.42	0.007365	91,036	23.53
70	0.027295	71,108	13.30	0.018160	81,235	15.69
80	0.068216	45,986	7.62	0.047669	60,540	9.16
90	0.184140	14,091	3.72	0.143437	25,512	4.52
100	0.383713	543	1.94	0.325102	1,981	2.29
110	0.625029	1	1.06	0.582207	7	1.16
119	0.969625	0	0.53	0.969625	0	0.53

[a] Probability of dying within one year.
[b] Number of survivors out of 100,000 born alive.

Life expectancy is the average number of years of life remaining at a given age. It is the average expected lifespan of an individual.

From the table above, one can see the death probability is very small at early ages (usually between 0 and 16 years), and then increases exponentially as shown in Figure (48).

Figure (48): Exponential increase in death probability and exponential decrease in life expectancy with increasing age

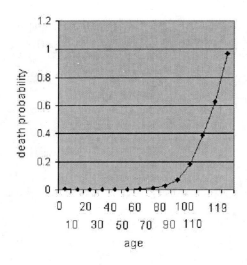 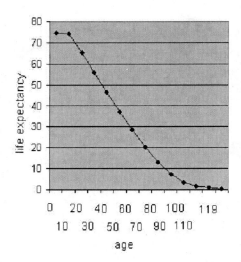

Now, let's assign

P_x for probability of surviving from age 0 to age x,   and

N_x for average number of offspring produced by and individual of age x.

In the curve of above figure, Figure (48), there are two portions of the curves: continuous and discrete (exponential).

For the continuous curve, one can consider the following Fisher's reproductive value:

$$V_x = \frac{\sum_{y=x}^{\infty} P_y N_y}{\sum_{y=0}^{\infty} P_y N_y}$$

If the curve is discrete, the value would be:

$$\frac{\int_{y=x}^{\infty} P_y N_y \, d_y}{\int_0^{\infty} P_x N_x \, d_x}$$

Which is the net reproductive rate of the population.

The average age of a reproducing adult in the continuous period is the generation time

$$\sum_{y=0}^{\infty} P_y N_y$$

The average age of a reproducing adult in the discrete period is the generation time in discrete form

$$\int_0^{\infty} P_x N_x d_x$$

Fisher's reproductive value is not accurate since it has a flaw in its mathematics due to his neglect of changes in population characteristics under natural selection; with the result that N equals the logarithmic rate of change of the total reproductive value of a population. A modified definition is given that does have this property. However, for practical uses in population genetics it may be better to follow Fisher's original definitions, despite the flaws in them. The flaw is how to provide a mathematically rigorous exposition of how optimization ideas can be applied to the operation of natural selection when the population is divided into classes. For optimization, Px Nx dx should be equal to zero. This can happen only when the age equal to zero (first day of birth) or number of offspring equals to zero. Optimization of 0 can not be calculated.

3-17 Basal metabolic rate

Basal metabolic rate, or BMR, is the minimum calorific requirement needed to sustain life in a resting individual in environment with a neutral temperature . It can be looked at as being the amount of energy (measured in calories) expended by the body to remain in bed asleep all day. For example, if your BMR is 2000 calories per day, then eating more than 2000 calories a day will result in weight gain. Eating less than 2000 calories per day will cause weight loss. BMR depends on the level of activity – the more active we are, the more energy we will need. In simple terms, if we have a faster metabolism we burn calories more efficiently and store less fat. If we have a slower metabolism we burn calories less efficiently and therefore store more calories as fat.

BMR is sufficient only for the functioning of the vital organs, such as the heart brain, lungs, liver, kidneys, muscles, skin, sex organs and the nervous system.

BMR is calculated using body surface area (calculated from height and weight), age, and sex. BMR can be responsible for burning up to 70% of the total calories expended, but this figure varies due to the following factors:

- Gender: Men have a greater muscle mass and a lower body fat percentage. This means they have a higher basal metabolic rate.
- Genetics: Some individuals are born with slower metabolisms; some with faster metabolisms.

- Body Surface Area: This is a reflection of height and weight. The greater the Body Surface Area factor, the higher is the BMR.
- Adipose tissues (fat) Percentage: The lower your body fat percentage, the higher your BMR. Men usually have lower adipose percentage than women, and this is one reason why men generally have a 10-15% higher BMR than women.
- Diet: Starvation can dramatically reduce BMR.
- Body Temperature/disease: For every increase of 0.5C in internal temperature of the body, the BMR increases by about 7 percent. So a patient with a fever of 41C (about 4C above normal) would have an increase of more than 50 percent in BMR.
- External temperature: External temperature affects the BMR. Exposure to cold temperature causes an increase in the BMR, so as to compensate for the extra heat needed to maintain the body's internal temperature.
- Endocrine system: The endocrine system is involved in regulating metabolism, growth, development, puberty, tissue function, internal environment (temperature, water balance, and ions) and also plays a part in determining mood. For example, thyroxin (produced by the thyroid gland) is a key BMR-regulator which speeds up the metabolic activity of the body. The more thyroxin produced, the higher the BMR. If too much thyroxin is produced (a condition known as thyrotoxicosis) BMR can actually double. If too little thyroxin is produced (myxoedema) BMR may shrink to 30-40 percent of normal. Like thyroxin, adrenaline also increases the BMR but to a lesser extent. Other hormones such as FSH (follicle stimulating hormone), LH (leutenizing hormone), insulin, glucagon, adrenaline, androgen, estrogen, melatonin, ADH (antidiuretic hormone which regulates water excretion by the kidneys) and others also regulate the BMR.
- Physical exercise: Physical exercise burns more calories, and increase your BMR to build more lean tissues over fat tissues. Physical exercise makes you burn more calories even when sleeping.

3-17-1 Calculation formulas of BMR

The original equations for calculating the BMR for men and women are:

$$BMR = (13.7516 \times W + 5.0033 \times H + 6.755 \times A + 66.473) \quad \text{kcal/day for men}$$

$$BMR = (9.5634 \times w + 1.8496 \times H + 4.6756 \times A + 655.0955) \quad \text{kcal/day for women}$$

W = weight (kg), H = height (cm), A = age (year)

The most recent equation is:

$$BMR = (9.99 \times W + 6.25 \times H + 4.92 \times A + S)$$

Where S = +5 for men, and = -161 for women

Example: A man of 170 cm height, 80 kg weight, and 50 years old would need 2112.7 kcal/day. A woman of the same weight, height and age would need 1946.7 kcal/day.

The above formulae are based on body weight, height and age, but do not take into account the difference in metabolic activity between lean body mass and body fat. Sometimes, BMR for women of same weight, height and age is more than men due to the accumulation of adipose tissues (fat) in women.

A more accurate and usable formula is the following formula based on lean body mass:

BMR = 370 + (21.6 x LBM)

Where LBM is the lean body mass in kg.

3-17-2 Break down of BMR in humans

About 70% of a human's total energy expenditure is due to the basal life processes within the organs of the body (see table). About 20% of one's energy expenditure comes from physical activity and another 10% from thermogenesis, or digestion of food (postprandial thermogenesis), see http://en.wikipedia.org/wiki/Basal_metabolic_rate

Table (6) shows the break down of BMR.

Table (6): Percentage of BMR in body's organs

Breakdown of BMR	Percentage%
liver	27
brain	19
heart	7
kidneys	10
skeletal muscle	18
other organs	19

3-17-3 BMR and balance between catabolism and anabolism

Catabolism and anabolism are two different states that describe whether the body is building tissue or breaking it down. The breakdown of large molecules into smaller molecules associated with release of energy is catabolism. The building up process is termed anabolism. The breakdown of proteins into amino acids is an example of catabolism while the formation of proteins from amino acids is an anabolic process. In other words, catabolism means that the body is breaking down tissue. Exercise, whether it's cardio or weightlifting, causes tiny tears in your muscle. The longer and harder you exercise, the more damage you'll cause to your muscle tissue. Anabolism means that your body is building or repairing tissue. Resting repairs your damaged muscle tissue and put on size and weight.

Let's see how molecules of sugar, fat, and protein break down into smaller molecules, releasing energy.

3-18 Sugar, fat and protein

Sugar (glucose) has molecular formula $C_6H_{12}O_6$. Oxidizing carbon means burning it. For example, take one molecule of methane $CH4$ and oxidize it as shown below:

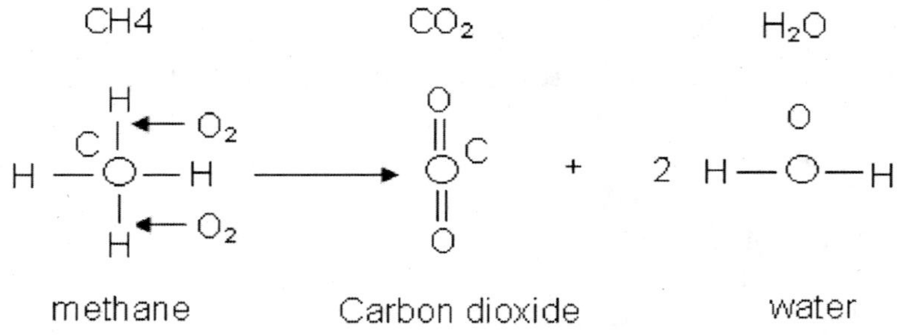

methane   Carbon dioxide   water

In oxidation, the carbon of the methane is combined with oxygen (CO_2), and the hydrogen of the methane is combined with oxygen. This is because oxygen loves hydrogen (oxygen ends with 6 electrons, and carbon ends with 4 electrons).

Sugar oxidation separates carbons from the sugar chain and combines them with oxygen, and oxygen combines with hydrogen as illustrated in Figure (49). The product is carbon dioxide and water.

Figure (49): Oxidation of glucose

The equation of sugar burning is:

$$C_6H_{12}O_6 + 6O_2 \longrightarrow 6CO_2 + 6H_2O$$

From Figure (49), it can be seen that you need an equal proportions of oxygen and carbon dioxide in order to burn the sugar you consume. So the percentage of respiratory between oxygen and carbon dioxide equals 1, or:

$6O_2 / 6CO_2 = 1$ (for sugar)

The same analysis can be applied to fat and protein which have the following respiratory percentages:

$O2 / CO2 = 1.4$ (for fat)

$O2 / CO2 = 1.25$ (for protein)

Therefore, the lung needs more oxygen input than Carbon dioxide output in order to burn fat and protein, Figure (50).

Figure (50): Oxygen and carbon dioxide for burning sugar, fat and protein

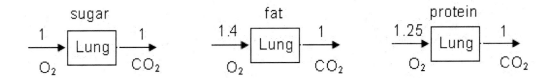

3-18-1 Glycation

Glycation reaction was discovered by the French chemist L. Millard in 1912 while studying foods. Sugar reacts non-enzymitically with a wide range of proteins to form early glycation products, which are called fructosamine products. So sugars react with protein amino groups to form a variety of protein-sugar compounds accumulates in vivo on vascular wall collage and basement membranes a function of age and levels of glycemia.

Glycation is a process by which proteins, certain fats, and glucose tangle (cross linked) together. It affects all body tissues, and tends to make them stiff and inflexible. Glycation causes most problems for organs where flexibility is most important, such as the heart, kidneys, skin and eyes.

Sugar molecules in the blood and in the cells chemically bond to proteins and to DNA. (This bonding is called "glycation", "the Maillard reaction", "the browning reaction", or "nonenzymatic glycosylation"). Over time, the sugar particles bound to the glycated proteins are chemically modified to become molecular structures called Advanced Glycation Endproducts (A.G.E.s), which is very reactive species. AGEs may be formed external to the body (exogenously) by heating (e.g., cooking or working at sugar factories) sugars with fats or proteins; or inside the body (endogenously) through normal metabolism and aging. Under certain pathologic conditions (e.g., oxidative stress due to hyperglycemia in patients with diabetes, and radical oxygen), AGE formation can be increased beyond normal levels. AGEs are now known to play a role as proinflammatory mediators in gestational diabetes as well, see
http://en.wikipedia.org/wiki/Advanced_glycation_end_product

Glycation and crosslinking have been implicated as strong contributors to many progressive diseases of aging, including vascular diseases (such as atherosclerosis, systolic hypertension, pulmonary hypertension, and poor capillary circulation), kidney disease, urinary incontinence, erectile dysfunction, stiffness of joints and skin, arthritis, cataracts, retinopathy, neuropathy, Alzheimer's Dementia, impaired wound healing, complications of diabetes, and cardiomyopathies (such as diastolic dysfunction, left ventricular hypertrophy, congestive heart failure and irregular heartbeats (arrhythmia) and sudden cardiac arrest).

In collagen (the most abundant protein in the body) AGEs form covalent, intermolecular bonds. The accumulation of AGEs with plasma protein such as albumin, low density lipoprotein (LDL), immunoglubin G (IgG) and nitrogen oxide is trapped in the basement membrane (endothelium). The following vascular pathology may occur:

Albumin + AEGs → protein accumulation which thickens the BM

LDL + AEGs → Trapping LDL in artery for oxidation (toxification)

IgG + AEGs → activation of inflammation

Nitrogen oxide + AEGs → Quenching of NO to cause poor vasodilation (widening blood vessels)

In the market, there are some drugs used as inhibitors of glycation crosslinking such as Pimagedine and Carnosine. Aspirin may also inhibit the formation of pathological A.G.E. crosslinks. For example, chronic users of aspirin have fewer cataracts.

3-19 BMR and aging

BMR is the largest component of energy expenditure and comprises about 70% of total expenditure in most adults. A decline in BMR with aging is well recognized. A study by Keys et al. documented a decline in BMR with age of 1–2% per decade; based on this assessment, a reduction in BMR of ~400 kJ/day can be predicted between 20 and 70 years of age. This assumption by Keys et al is disputed that the relation between BMR and years of age is a straight line.

However, Kleiber's law, named after Max Kleiber's biological work in the early 1930s, is the observation that, for the vast majority of animals, the animal metabolic rate is equal to the power ¾ of the animal weight. Therefore, the BMR and age is exponential as per Kleiber's law, Figure (51). Thus a cat, having a mass 100 times that of a mouse, will have a metabolism roughly 31 times greater than that of a mouse. In plants, the metabolic rate equals to the plant mass (weight equals mass multiplied by gravity). The exponent for Kleiber's law, was a matter of dispute for many decades. It is still contested by a diminishing number as being ⅔ rather than the more widely accepted ¾.

Figure (51): Kleiper' law showing exponential relationship between age and metabolic rate

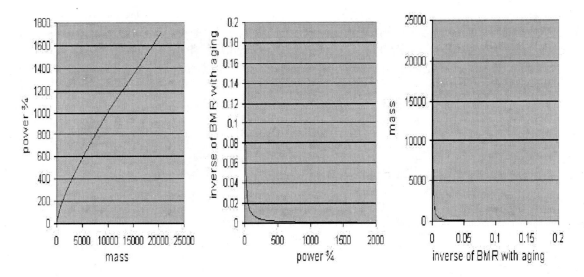

BMR is the rate at which oxygen is consumed and carbon dioxide is produced by the organism under states of physical and mental rest, 12 to 14 hrs after the last meal and in neutral ambient temperature.

BMR exponentially decreases with aging, but the amount of the fall varies with the study and the criterion of the measurement (body mass vs. lean body mass). BMR is normally computed in terms of oxygen consumption in kilocalories per unit of body surface and time. With aging from adolescence to senescence, metabolic rate decreases exponentially. In females values are always slightly lower than in men. The cause of such an increasing decline in BMR is not definitely known. It may be associated with ages due to:

- Decreased level of thyroid hormones, T3, with age
- Increased level in adipose (fat) body mass (less metabolically active) relative to lean body mass (more metabolically active
- Decreased metabolic rate of cytoplasmic masses with aging
- Decreased amount of potassium (the major electrolyte) with aging
- Decreased level in creatinine, a product of creatine which is the product of muscle metabolism.

3-20 Heat shock and osmotic stress

Heat shock and osmotic increase the level of sorbitol. Sorbitol, is a sugar alcohol which is metabolized slowly in the human body. Sorbitol is not only a sugar substitute, it is used as a humectant and thickener in many over the counter cosmetic products. Many transparent gels on the market contain sorbitol for its highly refractive index that is essential for many transparent formulations. Sorbitol is made through the reduction of glucose. This reduction of the glucose changes the aldehyde group to an additional hydroxyl group. This is where sorbitol gets the name sugar alcohol. Figure (52) shows the difference between sorbitol and glucose.

Figure (52): Chemical structure of sorbitol and glucose

glucose $C_6H_{12}O_6$

sugar alcohol (sorbitol) $C_6H_{14}O_6$

Both heat shock and osmotic stress between 200 and 300 mM (millimolar) of sorbitol clearly induced heat shock proteins 70 and 27 mRNA levels (70 mRNA, the heat shock protein (HSP70) gene is located in chromosome 14, it is now considered as a molecular 'chaperone' (a protein that assists folding/unfolding in molecular biology and a cell-protective agent). It may be closely related to the pathogenesis of dementia of the Alzheimer type, 27 mRNA is widely expressed in bones and kidneys). Heat shock and osmotic stress also affect the mitogen which is usually some form of a protein that encourages a cell to commence cell division, and activates protein kinase family of proteins which plays an important part in the coordination of gene responses to various stress conditions.

Exposure to heat, salinity, or osmotic stress resulted in alterations in cyanobacterial-protein syntheses. Three prominent types of modifications were noted: (i) synthesis of several proteins declined, (ii) synthesis of certain other proteins was selectively enhanced, and (iii) synthesis of a new set of proteins was induced de novo. Some of these responses were observed under all stress conditions (tentatively called the common-stress proteins), while others were found to be specific to heat (heat shock proteins) or salinity/osmotic stress (osmotic-stress proteins), see http://jb.asm.org/cgi/reprint/171/9/5187.pdf.

3-20-1 Effects of heat shock and osmotic stress on aging

Heat, hyperosmetic medium and ethanol are stressing factors. Since the CuZnSOD-deficient strain DSCD1-1C shows a significantly higher sensitivity to superoxide-generating agents, than the other strains tested, it was chosen for studying the protective effects of stress treatment with respect to subsequent exposure to increased intracellular superoxide flux. Previously we have shown that deficiencies in superoxide dismutases (SOD) result in an almost shortening of the life span. Experiments on yeast Saccharomyces cerevisiae showed that mutation in cells are known to be hypersensitive to various types of oxidative stress which can be developed due to heat shocks and osmotic stresses. Heat shocks and osmotic stresses can also reduce the number of

294

enzymes which constitute the first line of defense against oxidative stress, and lead to increased levels of reactive oxygen species which result in intracellular damage or even in cell death. Figure (53) illustrates the effect of heat shock and osmotic stress on survival, see http://www.actabp.pl/pdf/2_2000/355-364s.pdf.

There are two main changes in osmotic stress; smooth changes and step changes. If the osmotic pressure changes from hypotonic to hypertonic in steps, the survival rate would be decrease progressively.

Figure (53): Effect of heat shock and osmotic stress on survival

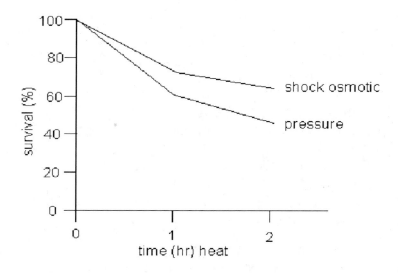

Devoid of CuZnSOD in atmospheric 100% oxygen and in the presence of menadione

3-21 Gene therapy

Gene therapy is an experimental technique that uses genes to treat or prevent disease. In the future, this technique may allow doctors to reverse aging by introducing a gene into a aged individuals to rejuvenate cells. Researchers are testing several approaches to gene therapy, including:

- Replacing a mutated gene that causes disease with a healthy copy of the gene.
- Replacing old genome with young ones.
- Inactivating, or detaching a mutated gene that is malfunctioning.
- Introducing a new gene into the body to help fight a disease.

Genes, which are carried on chromosomes, are the basic physical and functional units of heredity. There are two types of genes: Germ line gene, i.e., sperm or eggs, are modified by the introduction of functional genes, which are ordinarily integrated into their genomes. Therefore, the change due to therapy would be heritable and would be passed on to later generations. The Somatic gene therapy transfers the therapeutic genes into the somatic cells of a patient. Any modifications and effects will be restricted to the individual patient only, and will not be inherited by the patient's offspring.

This new method, theoretically, should be highly effective in counteracting genetic disorders and hereditary diseases. However, many jurisdictions and ethical principles prohibit this from exercising in human beings, for a variety of technical and ethical reasons. Genes are specific sequences of bases (nucleotides) that program instructions on how to synthesize enzymes and proteins. Although genes get a lot of attention, it's the proteins that perform most life functions and even make up the majority of cellular structures. When genes are altered so that the encoded enzymes and proteins are unable to perform their normal functions, diseases and aging are the consequences.

Gene therapy is being thoroughly researched as a cure for several genetic diseases. Out of all the genetic disorders, gene therapy for both sickle cell and hemophilia diseases has the most favorable characteristics for this potential cure. Gene therapy works in hemophilia by using DNA as the drug and viruses as the deliverer. For example, a virus containing the gene that produces Factor VIII or Factor IV (in case of Hemophilia B) is injected into a large group of cells in the patient. The hope of gene therapy is to have the cell produce more of the cured cells and spread them throughout the rest of the body. If successful, the patient would never need factor replacement therapy again and would be cured of diseases. Figure (54) indicates the process of gene therapy injection.

Figure (54): Process of gene therapy

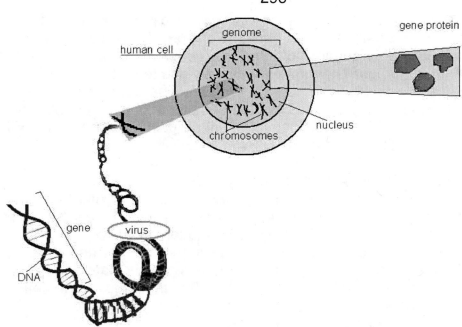

Los Angeles times, October 30, 2009, quoted "Pennsylvania researchers using gene therapy have made significant improvements in vision in 12 patients with a rare inherited visual defect, a finding that suggests it may be possible to produce

similar improvements in a much larger number of patients with retinitis pigmentosa and macular degeneration"

3-21-1 Gene therapy and aging

Age-related deterioration in critical brain networks may be restored by gene therapy, according to a study in monkeys presented at the American Academy of Neurology's 52nd Annual Meeting in San Diego, CA, April 29 -- May 6, 2000. This finding lends support to treat Alzheimer's disease using a similar gene therapy approach, say the study's authors. Researchers from the University of California in San Diego found that normal aging in monkeys causes a 28 percent decline in the density of certain brain networks originating from nerve cells called neurons deep in the brain.

The scientists discovered that they were able to restore these connections by transplanting brain cells genetically programmed to release a protein called "nerve growth factor." "It would be inappropriate to suggest that this approach could be used to treat the course of normal aging in all organs, but it is not a far stretch to suggest that this may be useful in the treatment of Alzheimer's disease.

Other studies performed on various experimental model systems indicate that gene therapy can increase longevity and slow aging, even if in laboratories.

Generally, such procedure required sophisticated technical processes. Overexpression of some genes, such as stress response and antioxidant genes, in some model systems also extends their longevity.

Studies have also shown that the effects of adding one or multiple copies of various genes, that leads to the increased expression of their gene products, has resulted in the extension of the life span in model systems such as worms, fruitflies, rodents and cultured cells. Some such transgenic manipulations include the addition of gene(s) such as antioxidant genes superoxide dismutase (SOD) and catalase, NAD+-dependent histone deacetylases sirtuins, forkhead transcription factor FOXO, heat–shock proteins (HSP), heat–shock factor, protein repair methyltransferases and klotho, which is an inhibitor of insulin and IGF-1 signaling. Another system in which genetic interventions have been tested is the Hayflick system of the limited proliferative life span of normal diploid differentiated cells in culture.

3-21-2 Hayflick limit and gene therapy

Leonard Hayflick, PhD, a professor of anatomy at the University of California, San Francisco is best known for his aging theory known as the Hayflick Limit. It places the maximum potential lifespan of humans at 120, the time at which too many cells can no longer split and divide to keep things going. This means that most of human cells stop dividing at the age of 120 years.

Hayflick said "aging occurs because the complex biological molecules of which we are all composed become dysfunctional over time as the energy necessary to keep them structurally sound diminishes. Thus, our molecules must be repaired or replaced frequently by our own extensive repair systems." The question is that "can gene therapy be used to repair our aged molecules and cells"? Extensive studies showed that gene therapy can extend the life span of prokaryotic species.

Hayflick added in his July speech at the World Congress of Gerontology and Geriatric in Paris "These repair systems, which are also composed of complex molecules, eventually, suffer the same molecular dysfunction. The time when the balance shifts in favor of the accumulation of dysfunctional molecules is determined by natural selection — and leads to the manifestation of age changes that we recognize are characteristic of an old person or animal. It must occur after both reach reproductive maturity, otherwise the species would vanish." He continued to say "these fundamental molecular dysfunctional events lead to an increase in vulnerability to age-associated disease. Therefore, the study, and even the resolution of age-associated diseases, will tell us little about the fundamental processes of aging".

To conclude, Hayflick pointed out that some types of cells, such as those that produce red and white blood corpuscles, can divide millions of times. Others, such as most nerve cells, do not reproduce at all. If a cell's Hayflick limit is 50, for example, it will divide 50 times and then become senescent. It withers and dies. When enough of our cells die, we die.

In 1961, Leonard Hayflick and Paul Moorhead discovered that human cells derived from embryonic tissues can only divide a finite number of times in culture. They divided the stages of cell culture in three phases: Phase I is the primary culture, when cells from the explants simply multiply to cover the surface of the culture flask. Phase II represents the period when cells divide in culture. Briefly, once cells cover a flask's surface, they stop multiplying. For cell growth to continue, the cells must be sub cultivated. To do so, one removes the culture's medium and adds a digestive enzyme called trypsin that dissolves the substances keeping cells together. If you add growth medium afterwards, you obtain the cells in suspension that can then be divided by two--or more--new flasks. Later, cells attach to the flask's floor and start dividing once again until new sub cultivation is required. Cells divide vigorously and can often be subcultivated in a matter of a few days. Eventually, however, cells start dividing slower, which marks the beginning of Phase III. Eventually they stop dividing at all and may or not die. Hayflick and Moorhead noticed that cultures stopped dividing after an average of fifty cumulative population doublings. This phenomenon is known as Hayflick's limit, Phase III phenomenon, or, as it will be called herein, replicative senescence as per Figure (55).

Figure (55): Phases of replication of cell culture

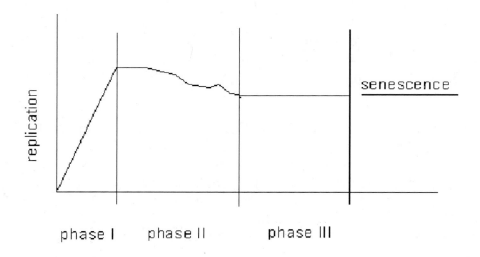

Hayflick and Moorhead worked with fibroblasts, a type of cell found in connective tissue; produces collagen. RS (replicative senescence) has been found in other cell types: keratinocytes (epidermal cells), endothelial cells (endothelium cells are the thin layers of cells that line the interior surface of blood vessels), lymphocytes (a lymphocyte is a type of white blood cell in the immune system), adrenocortical cells, vascular smooth muscle cells, chondrocytes (are the only cells found in cartilage), etc. In addition, RS is observed in cells derived from embryonic tissues, in cells from adults of all ages, and in cells taken from many animals: mice, chickens, Galapagos tortoises, etc. Early results suggested a relation between the number of CPDs (cumulative population doublings) cells undergo in culture and the longevity of the species from which the cells were derived. For example, cells from the Galapagos tortoise, which can live over a century, divide about 110 times, while mouse cells divide roughly 15 times. In addition, cells taken from patients with progeroid syndromes such as Werner syndrome endure far less CPDs than normal cells. Exceptions exist and certain cell lines never reach RS. Some cells have no Hayflick limit. Barring trauma from outside, they are immortal. They can be killed, but they do not age. The "lowly" bacteria are immortal. They can be killed – by heat, starvation, radiation, lack of water, or being eaten by another organism. But they do not age. Bacteria keep on dividing forever, until some outside agency kills them. Cancer cells are similarly immortal. They keep on dividing and dividing, endlessly, unless they are killed or their host dies. "HeLa" cells, taken from the tumor of Henrietta Lacks in 1951 (Henrietta Lacks (August 18, 1920 – October 4, 1951) was the involuntary donor of cells from her cancerous tumor, which were cultured by George Otto Gey to create an immortal cell line for medical research. This is now known as the Hela cell line), are still reproducing as vigorously as they did nearly 50 years ago. Human germline cells -- ova and sperm cells -- also show no Hayflick limit.

Senescent cells are growths arrested in the transition from phase G1 to phase S of the cell cycle. The growth arrest in RS is irreversible in the sense that growth factors cannot stimulate the cells to divide, even though senescent cells can remain metabolically active for long periods of time.

Recent studies showed that marrow cells (Marrow stromal cells, MSCs) may be used to repair senescence cells and gene therapy. If marrow cells are to be used for cell and gene therapy, it will be important to define the conditions for isolation and expansion of the cells. As demonstrated by Friedenstein and colleagues, MSCs are relatively easy to isolate from marrow from most species by their adherence to tissue culture plates and flasks. However, the cells display several unusual features as they expand in culture. The difficulty of carrying out experiments in animal models with MSCs and other marrow cells has prompted scientists to develop a coculture system to study the repair of injured cells and tissues by MSCs. In initial experiments, MSCs were cocultured with heat-shocked human small airway epithelial cells. Figure (56) is a schematic showing MSCs with gene green fluorecent protein injected into senescence cells to produce repaired cells.

Figure (56): Using MSCs as gene therapy

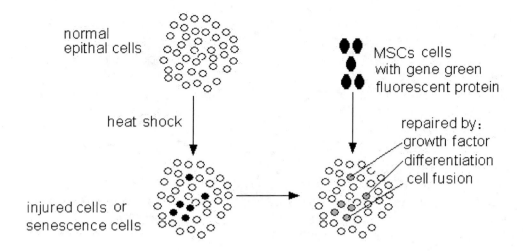

3-22 Polymorphism genetics and aging

Polymorphism is the coexistence of several distinctly different types in a population (groups of animals of one species). Examples include the different blood groups in humans, different colour forms in some butterflies, and snail shell size, length, shape, and colour.

On the evolutionary background and on the basis of data collected in the last 10 years on centenarians, it is suggested that genes involved in immune responses and inflammation play a major role in human ageing and longevity. Moreover, variants of gene involved in energy production, oxygenation stress, IGF1/insulin pathway and mitochondrial DNA (mtDNA) gem line inherited polymorphism and mutations are also major actor in longevity and aging. Genetic polymorphisms are different forms of a DNA sequence. "Poly" means many, and "morph" means form or shape. Polymorphisms are a type of genetic diversity within a population's gene pool. They can be used to map (locate) genes such as those causing a disease, and they can help match two samples of DNA to determine if they come from the same source. Polymorphism has many uses in medicine, biological research, and law enforcement). Depending on its exact nature, a polymorphism may or may not affect biological function. The following are the major issues related to genetic polymorphism:

- The human body has been naturally selected for and shaped in order to survive until the age of reproduction and the average life expectancy at birth was quite short until a couple of centuries ago (Life expectancy did not exceed 40-45 years until a couple of centuries ago even in the most developed countries).
- The immune system has also been formed by evolution and it is quite efficient in coping with acute infections in young people, but probably not in aged people.

- Polymorphisms are very important factors involved in the immune system and thus they are of different influence and responses at different years of aging.
- Polymorphisms affect the pleiotropy (pleiotropy occurs when a single gene influences multiple phenotypic traits). Phenotypic traits represent diverse coloration and patterning in shapes and appearances. Consequently, a new mutation in the gene may have an effect on some or all traits simultaneously. The change in traits is probably due to the change in transcription from DNA to RNA and the translation from RNA to protein or vise versa, Figure (57).

Figure (57): Change in transcription from DNA to protein

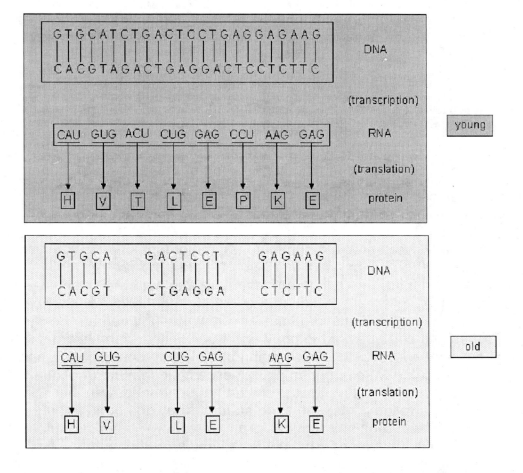

- Alterations of signalling pathways between RNAs and proteins and their effects on other pathways such as the IGF1/insulin pathway and mitochondrial DNA are also emerging as critical target for the development of frailty and major age-related diseases.
- There are hundreds of thousands of SNP (single nucleotide polymorphisms) loci throughout the human genome, making them causes for aging and shortened life spans. When the sequences of

nucleotides are changed, the sequence of protein will be changed, and the result is the change in the sequence of amino acids which make the alleles. The change in the positions of the alleles and thus the genes on a chromosome is its locus (plural, loci).

- Polymorphism could have an effect on many proteins. For example, apoliprotein E is an important determinant of blood levels, atherosclerosis, and longevity. The allele of the apoliprotein E is inversely correlated with human longevity (Castro et al., 1999), and individuals with this allele have doubled risk of Alzheimer's disease (Havlik et al., 2000).
- Polymorphism can alter any protein in the human, even proteins which are responsible for the nervous system. Tau proteins are abundant in neurons of the central nervous system and are less common elsewhere. Abnormal tau hyperphosphorylation has been suggested as being one of the central events in the development of neurofibrillary

tangles, which are one of the characteristic neuropathological lesions found in Alzheimer's disease brains.

3-23 Accelerated aging diseases

Many diseases can accelerate aging. All of them are due to DNA repair-deficiency disorders. Some of these diseases are:

a) Progeria: Human progeria comes in two major forms, Werner's syndrome (adult-onset progeria) and Hutchinson-Gilford syndrome (juvenile-onset progeria). There is no full agreement as to whether or not progeria is a cause of aging. Most clinicians believe that progeria is truly a form of early aging, although only a segmental form in which only certain specific tissues and cell types of the body age early. Hutchinson-Gilford children show what appears to be early aging of their skin, bones, joints, and cardiovascular system, but do not senesce their immune or central nervous systems. Clinically, the children appear aged, with thin skin, baldness, swollen joints, and short stature. They do not go through puberty. The face is strikingly old in appearance. The typical Hutchinson-Gilford child looks more like a centenarian than like other children, and may look more like other progeric children than like members of their own families.

b) Ataxia telangiectasia (A-T): Ataxia-telangiectasia is a rare, childhood neurological disorder that causes degeneration in the part of the brain that controls motor movements and speech. The first signs of the disease, which include delayed development of motor skills, poor balance, and slurred speech, usually occur during the first decade of life. Ataxia telangiectasia affects the cerebellum (the body's motor coordination control center) and also weakens the immune system in about 70% of the cases, leading to respiratory disorders and anincreased risk of cancer, mainly malignant lymphoma at early age.

Ataxia telangiectasia (A-T) is a rare autosomal recessive multisystem disease. The disease could be due to an increase in alpha-fetoprotein (AFP) level which reduces the IgG protein family. Friedreich's Ataxia (results from the degeneration of nerve tissue in the spinal cord) is the most common inherited ataxia, with symptoms generally appearing between the ages of 8 and 15.

c) Bloom syndrome: It was discovered and first described by dermatologist Dr. David Bloom in 1954. Bloom syndrome is a rare autosomal recessive chromosomal disorder characterized by a high frequency of breaks and rearrangements in an affected person's chromosomes, and characterized by telangiectases (related to the presence of venous hypertension within underlying varicose and spider veins) and photosensitivity. People with Bloom syndrome are much smaller than average, and often have a high-pitched voice. Characteristic facial features include a long, narrow face; small lower jaw; and prominent nose and ears. They tend to develop pigmentation changes and dilated blood vessels in the skin, particularly in response to sun exposure. These changes often appear as a butterfly-shaped patch of reddened skin on the face. The skin changes may also affect the hands and arms.

d) Cockayne syndrome: It is named after English physician Edward Alfred Cockayne (1880-1956). Cockayne syndrome is a rare inherited disorder in which people are sensitive to sunlight, have short stature, and have the appearance of premature aging. In the classical form of Cockayne syndrome (Type I), the symptoms are progressive and are characterized by normal fetal growth with the onset of abnormalities in the first two years of life. An early onset or congenital form of Cockayne syndrome (Type II) is apparent at birth and involves very little neurological development after birth. Death usually occurs by age 7. Interestingly, unlike other DNA repair diseases, Cockayne syndrome is not linked to cancer. Other possible signs and symptoms include hearing loss, eye abnormalities, severe tooth decay, and problems with internal organs. Figure (58) shows a patient with Cockayne syndrome.

Figure (58): picture of a patient with Cockayne syndrome

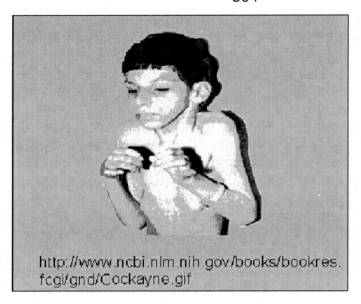

http://www.ncbi.nlm.nih.gov/books/bookres.
fcgi/gnd/Cockayne.gif

e) Xeroderma pigmentosa (XP): Xeroderma pigmentosa is a rare condition passed down through families in which the skin and tissue covering the eye are extremely sensitive to ultraviolet light. It is caused by the ultraviolet light -- such as that found in sunlight – which damages the genetic material (DNA) in skin cells. Normally, the body repairs this damage. But in persons with xeroderma pigmentosa, the body does not fix the damage. As a result, the skin gets very thin and patches of varying color (spotty pigmentation) appear. This disorder leads to multiple basal cell carcinomas (basaliomas) and other skin malignancies at a young age. In severe cases, it is necessary to avoid sunlight completely. The two most common causes of death for XP victims are metastatic malignant melanoma and squamous cell carcinoma. XP is about six times more common in Japanese people than in other groups. A picture of a patient with xeroderma pigmentosa is shown in Figure (59).

Figure (59): Patient with xeroderma pigmentosa

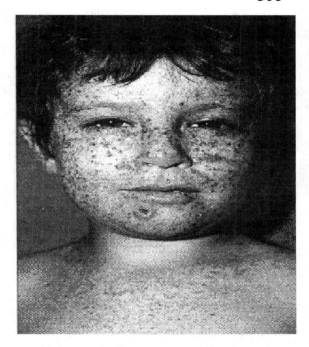

http://genetica.zip.net/images/Xeroderma.jpg

f) Rothmund–Thomson syndrome (R-T): R-T is a rare autosomal rece3ssive skin condition originally described by August Von Ruthmond (1830-1906). In 1868. Matthew Sydney Thomson (1894-1969) published further descriptions in 1936. Rothmund-Thomson syndrome is a rare condition that affects many parts of the body, particularly the skin. People with this condition typically develop redness on the cheeks between ages 3 months and 6 months. Over time the rash spreads to the arms and legs, causing patchy changes in skin coloring, areas of skin tissue degeneration (atrophy), and small clusters of enlarged blood vessels just under the skin (telangiectases). These skin problems persist for life, and are collectively known as poikiloderma. Figure (60) shows a patient with Rothmund–Thomson syndrome.

Figure (60): A patient with Rothmund–Thomson syndrome

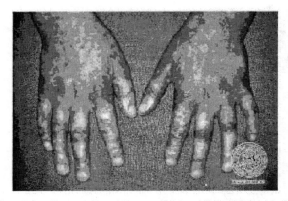

g) Trichothiodystrophy:

Trichothiodystrophy IBIDS syndrome (also known as "IBIDS syndrome " and "Tay's syndrome") was first described by Tay Chong Hai, a Singaporean doctor who discovered it in 1971. Patients with trichothiodystrophy have abnormally sulfur deficient brittle hair and nails (because of reduced content of cysteine-rich matrix proteins which is a matrix-associated protein that elicits changes in cell shape, inhibits cell-cycle progression, and influences the synthesis of extracellular matrix), ichthyotic skin, and physical and mental retardation. Figure (61) shows a picture of a patient with trichothiodystrophy.

Figure (61): A patient with trichthiodystrophy

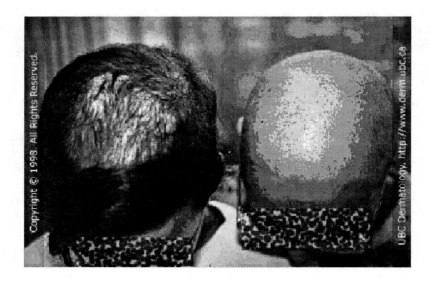

307

h)   Werner Syndrome: This was discussed earlier in this chapter.

3-24 Mitohormesis (Mitochondrial Hormesis)

In toxicology, hormesis is a dose response phenomenon characterized by a low dose stimulation, high dose inhibition, resulting in either a J-shaped or an inverted U-shaped dose response. Such environmental factors that would seem to produce positive responses have also been termed "eustress".

Hormesis may also be induced by endogenously produced, potentially toxic agents. For example, mitochondria consume oxygen which generates free radicals (reactive oxygen species) as an inevitable by-product. It was previously proposed on a hypothetical basis that such free radicals may induce an endogenous response cumulating in increased defense capacity against exogenous radicals (and possibly other toxic compounds). Recent experimental evidence from Michael Ristow's's laboratory strongly suggests that this is indeed the case, and that such induction of endogenous free radical production extends life span of a model organism. Most importantly, this induction of life span is prevented by antioxidants, providing direct evidence that toxic radicals may mitohormetically exert life extending and health promoting effects. In the metabolism inside the mitochondrion, a surprising new twist has emerged: Reactive oxygen species, derived from the mitochondrial electron transport system, may be necessary triggering elements for a sequence of events that result in benefits ranging from the transiently cytoprotective to organismal-level longevity. In toxicology, hormesis is a dose response phenomenon characterized by low dose stimulation, high dose inhibition, resulting in either a J-shaped or an inverted U-shaped dose response, Figure (62).

Figure (62): Reaction of mitochondrion during small and large doses of radicals

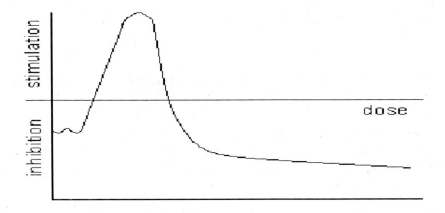

So if radicals increased beyond a limit, a signal from mitochondria is sent back to reduce radicals, Figure (63).

Figure (63): Effect of Mitohormesis

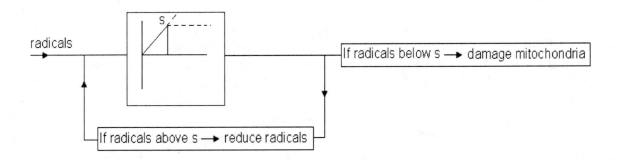

It is assumed that a low dose of a toxin may damage mitochondria, but after a certain level of toxin, mitochondria trigger certain repair mechanisms in the body, and these mechanisms, having been initiated, are efficient enough that they not only neutralize the toxin's effect, but even repair other defects not caused by the toxin.

However, whether hormesis is common or important is controversial, especially when applied to the field of ionizing radiation. The United States-based National Council on Radiation protection and Measurements concludes that the chance of causing cancer is always linearly proportional to the amount of radiation an individual receives, should be used at low doses too. The report squarely rejects almost all research showing radiation induced hormesis as being flawed in some way (e.g. the cancer a study focuses on does not exist in humans, a clear threshold could not be established in humans, the assumptions are seriously flawed, the hormetic effect is too short to be useful), see http://en.wikipedia.org/wiki/Hormesis

Linear radiation is called the Petkau effect. Dr. Abram Petkau at the Atomic Energy of Canada Whiteshell Nuclear Research Establishment, Manitoba and published in Health Physics March 1972.

Petkau had been measuring, in the usual way, the dose that would rupture a particular cell membrane. He found that 3500 rads delivered in 2¼ hours (26 rad/min) would do it. Then, almost by chance, he tried again with much weaker radiation and found that 0.7 rads delivered in 11½ hours (1 millirad/min) would also destroy the membrane. This was counter to the prevailing assumption of a linear relationship between total dose or dose rate and the consequences.

The radiation was of ionizing nature, and produced negative oxygen ions. Those ions were more damaging to the membrane in lower concentrations than higher (a somewhat counterintuitive result in itself) because in the latter, they more readily recombine with each other instead of interfering with the membrane. The ion concentration directly correlated with the radiation dose rate and the composition had nonmonotonic consequences.

2-25 Metabolic hormones affecting aging

3-25-1 Leptin aging

Leptin is produced by a specific gene found in fat cells called the obese (ob) gene. Small amounts of leptin are also secreted by cells in the epithelium, stomach and placenta. The amount of leptin found in people increases as their body fat increases. Leptin is an appetite suppressant. It stops you eating too much as well as makes you more active so you burn off more energy. Leptin also acts directly on the cells of the liver and skeletal muscles where it stimulates the oxidation of fatty acids in the mitochondria. This reduces the storage of fat in the mitochondria' tissues.

The common denominators that are found in almost all living beings which live the longest share the following traits:

- Reduction in glucose
- Low insulin levels
- Low leptin level
- Lower body temperature
- Low triglycerides
- Low percentage of body (visceral) fat
- Low thyroid levels

Most people in developed countries are overweight when they reach middle age. When they advance in age they exhibit a further increase in body weight, abdominal obesity, insulin resistance, and increased plasma leptin levels in proportion to their body mass, see http://biomed.gerontologyjournals.org/cgi/content/full/57/6/B225?ck=nck. The consequence of obesity in humans is decreased life span caused by an increase in all causes of death. Leptin's failure may be an important biological initiator of the events leading to obesity because the body is less sensitive to leptin. The failure of leptin injections to decrease body weight adequately in middle-aged obese and diabetic subjects demonstrates such a resistance. Taken together, these data, spanning humans and animal models, suggest that youth is a leptin-sensitive state, and that resistance to leptin occurs with aging. It is suggested that the failure of leptin to regulate food intake, body fat and its distribution, and insulin action occurs with advanced age. It also suggests that leptin resistance plays a major role in the metabolic syndrome that is typical of aging.

3-25-2 Resistin and aging

Resistin is poorly understood in humans as it has been primarily characterized in rats. It is, however, associated with insulin resistance at increased concentrations and is secreted more at higher BMIs (body mass index is a measure of body fat based on height and weight that applies to both adult men and women). Much of what is hypothesized about a resistin role in energy metabolism and T2DM (type II diabetes mellitus) can be derived from studies showing good correlations between resistin and obesity. Nevertheless, recent studies show lack of correlation between resistin and adiposity, while some investigators have reported that serum resistin concentrations are independently associated with risk for incident HF (heart failure), visceral fat and inflammation in older persons. In addition, plasma resistin levels correlated with triglycerides, waist conference, waist/hip ratio, systolic blood pressure, and ApoAI/ApoB ratio (ApoAI/ApoB ratio is the apolipoprotein (apo)B to apoA which is used to predict the risk of cardiovascular disease (CVD), similar to LDL/HDL cholesterol ratio).

3-25-3 Adiponectin and aging

Adiponectin is a protein hormone that modulates a number of metabolic processes, including glucose regulation and fatty acid catabolism. Adiponectin is exclusively secreted from adipose tissue into the bloodstream and is very abundant in plasma relative to many hormones. Levels of the hormone are inversely correlated with body fat percentage in adults. The hormone plays a role in the suppression of the metabolic derangements that may result in type II diabetes, obesity, atherosclerosis, non-alcoholic fatty liver disease (cirrhoses) and an independent risk factor for metabolic syndrome, see http://en.wikipedia.org/wiki/Adiponectin.

3-25-4 Ghrelin and Aging

Ghrelin is a hormone produced mainly in the enteroendocrine P/D1 cells of the gastric mucosa lining the fundus of the human stomach and epsilon cells of the pancreas that stimulates hunger. Ghrelin levels increase before meals and decrease after meals. It is considered the counterpart of the hormone leptin, produced by adipose tissue. Some surgeries, which are performed with very obese people to have a gastric bypass to lose weight end up with relatively little ghrelin, which may help explain why their appetites decrease after the surgery. Jules Hirsch, an obesity expert at Rockefeller University in Manhattan, said ghrelin clearly had a "profound effect on appetite". But, he added, "whether it will be useful in any way in the treatment of obesity remains to be seen".

3-25 Cytokines and aging

Cytokines are non-antibody proteins secreted by inflammatory leukocytes and some non-leukocytic cells, which act as intercellular mediators. Cytokines are small secreted proteins which mediate and regulate immunity, inflammation, and hematopoiesis. They differ from classical hormones in that they are produced by a number of tissue or cell type rather than by specialized glands. They generally act locally in a paracrine, intracrine or autocrine rather than endocrine system. Cytokine is a general name; other names include lymphokine (cytokines made by lymphocytes), monokine (cytokines made by monocytes), chemokine (cytokines with chemotactic activities), and interleukin (cytokines made by one leukocyte and acting on other leukocytes). Cytokines may act on the cells that secrete them (autocrine action), on nearby cells (paracrine action), or in some instances on distant cells (endocrine action).

Despite some controversial results, the available data is in favour of the hypothesis that pro-inflammatory cytokines play an important role in aging and longevity. Some studies proved that aging is associated with increased inflammatory activity reflected by increased circulating levels of TNF-α, IL-6, and other cytokines. TNF-α, and IL-6 are cytokines which cause tumor necrosis factors and are involved in systematic inflammation. TNF-α, and IL-6 are important in the regulation of immune cells and in the induction of apoptotic cell death, to induce inflammation, and to inhibit tumorigenesis and viral replication.

3-26 Polyphenol anti-aging

Polyphenol: A kind of chemical, characterized by the presence of more than one phenol unit per molecule, that (at least in theory) may protect against some common health disease and possibly certain effects of aging.

Polyphenols act as antioxidants. They protect cells and body chemicals against damage caused by free radicals such as O_2^-, H_2O_2, and OH^-, and reactive atoms that contribute to tissue damage in the body. For example, when low-density lipoprotein (LDL) cholesterol is oxidized (oxidation of LDL can be avoided by Vitamin E), it can be glued to arteries and cause coronary heart disease.

Polyphenols can also block the action of enzymes that cancers need for growth and they can deactivate substances that promote the growth of cancers. However, adverse effects of polyphenols have been evaluated primarily in experimental studies. It is known, for example, that certain polyphenols (flavone which has double bonded oxygen to the hexagonal molecule, Figure (64) may have carcinogenic/genotoxic effects or may interfere with thyroid hormone biosynthesis.

Figure (64): main types of phenols

gallic acid   cinnamic acid   flavone

Polyphenols are natural compounds widely found in plant foods. The most important dietary sources of polyphenols are onions, cocoa, tea, apples, and red wine, citrus fruit, berries and cherries and soy. Polyphenol anti-aging benefits have been extensively studied and it has been suggested that polyphenols are excellent antioxidants, even powerful than vitamins, which make them very effective agents in reducing cellular decay that typically happens with aging. Moreover, polyphenols are also known to posses excellent anti-inflammatory, anti-tumor, and anti-infection properties. They improve the immune response and reduce cancer-causing activities in the body (except for flavones). Polyphenols are also known to improve blood circulation and reduce the risk of heart diseases.

Phenol is used primarily for isolation and purification of DNA and RNA. Oxidized phenol can result in DNA damage, and cannot be used. The pH of phenol solutions dramatically changes the solubility of DNA. Oxidized phenol is effective at denaturing and precipitating most proteins, and the non-oxidized phenol is an effective means of purifying DNA or RNA from protein contaminants.

3-27 Epigenetic inheritance and aging

Epigenetic: Something that affects a cell, organ or individual without directly affecting its DNA. An epigenetic change may indirectly influence the expression of the genome.

Epigenetics has two meanings:

1) The study of heritable changes in gene function that occur without a change in the sequence of the DNA through:
   - Methylation of the DNA
   - Remodeling of chromatin
2) The study of a certain process which can occur in embryonic development through X inactivation (the inactivation of one X chromosome in females), and through gene silencing.

We shall deal with each of the above topics in detail.

3-28 Methylation of the DNA

In adult somatic tissues, DNA methylation typically occurs in a CpG dinucleotide context; non-CpG methylation is prevalent in embryonic stem cell. CpG, in genetics, a site where cytosine (C) lies next to guanine (G) in the DNA sequence. (The p indicates that C and G are connected by a phosphodiester bond). Methylation of DNA occurs at any CpG site. DNA methylation is a type of chemical modification of DNA that can be inherited and subsequently removed without changing the original DNA sequence. As such, it is part of the epigenetic code and is also the best characterized epigenetic mechanism. Because methylation is a common capability of all viruses for self non-self identification, the epigenetic code could be a persistent remnant of ancient viral infection events.

In general, when some compounds receive a methyl group, this "starts" a reaction (such as transcription of DNA into RNA, translating RNA into protein and turning a gene on or activating an enzyme). When the methyl group is "lost" or removed, the reaction stops (or protein is not produced, a gene is turned off or the enzyme is deactivated).

Some of the advantages of methylation reactions would be:

a. Methylation turns on a detoxification process that detoxifies the body of radicals.
b. Methylation turns on serotonin, and thus melatonin, production. This would improve your sleep.
c. Over-methyaltion could increase hyperness and aggression. So, too much methylation could increase mRNA and protein which stimulate energy and hyperness.
d. Methylation represses gene activity due to the corporation of P53 protein, which suppresses gene expression.
e. Methylation could damage the DNA due to the unexpressed and unregulated RNA, and thus could cause cancer, metabolic diseases, cardiovascular diseases, and autoimmune diseases.

Methylation occurs at bond between the cytosine and the guanine bases where the transcription of RNA takes place, Figure (65). In general, there are four mechanisms of methylations: methyltransferase (MTase) molecules progressively move 5' to 3' from their prime site, neighboring and cutting CpG dinucleotides, and diminished gene expression development.

Figure (65): Methylation of DNA

methylation

3-28-1 Remodeling of chromatin

Remodeling of chromatin is occurred due to the dynamic structural changes to the chromatin occurring throughout the cell division cycle. These changes range from the local changes necessary for transcriptional regulation to the overall changes necessary for chromatin segregation. Chromatin remodeling is an epigenetic phenomenon.

The importance of histones and chromatin structure (histone octamer plus DNA are called nucleosome) in the regulation of eukaryotic gene transcription has become much more widely accepted over the past few years. It has been clear for a decade that histones contribute to the regulation of transcription. More recent studies have led to the striking observation that several protein complexes involved in transcription regulation can function, at least in part, by modifying histones or altering chromatin structure. While it is clear that many of these protein complexes have functions in addition to chromatin modification, they show the importance of chromatin structure as a part of transcription regulation mechanisms, Figure (66).

Figure (66): Remodeling of chromatin for transcription

315

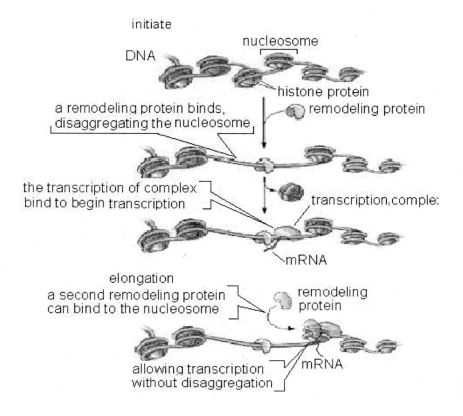

initiate

nucleosome

DNA

histone protein

a remodeling protein binds,
disaggregating the nucleosome

remodeling protein

the transcription of complex
bind to begin transcription

transcription complex

mRNA

elongation
a second remodeling protein
can bind to the nucleosome

remodeling protein

allowing transcription
without disaggregation

mRNA

Modified from: http://www.nature.com/scitable/content/Local-Remodeling-of-Chromatin-for-Transcription-4467

Activation and inactivation of gene transcription can not be completed without the linker histone H1 which represents the pillar of the chromatin. Studies have demonstrated that histones H1, H2A, H2B and H3 and H4 are effective mediators of transfection. Electron microscopy and biochemical studies have established that the bulk of the chromatin DNA is compacted into repeating structural units called nucleosomes. Each nucleosome contains about 200 base pairs (bp) of DNA associated with a histone octamer core consisting of two molecules each of histones H2A, H2B, H3, and H4, Figure (67). In addition, an association of many H1 histone molecules with the nucleosome is composing the chromatin.

Figure (67): Components of chromatin

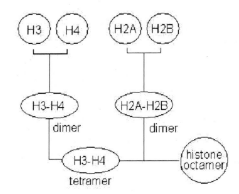

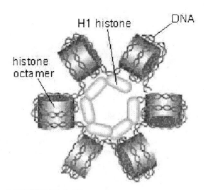

modified from:
http://www.mun.ca/biology/desmid/brian/BIOL2250/W
eek_Two/nucsoH1.jpg

3-28-2 Gene silencing

Gene silencing means interruption or suppression of the expression of a gene at transcriptional or translational levels. Scientists have been working on strategies to selectively turn off specific genes in diseased tissues for the past thirty years: Some techniques have been developed to control gene expression by turning them on or off at the DNA level, "because every disease starts at the level of malfunctioning gene expression, or viral or bacterial gene expression," said Dr. David Corey, professor of pharmacology and biochemistry at UT Southwestern Medical Center. In doing so, the research team may have paved the way for the development of new drugs designed to treat many serious diseases.

Gene silencing can be achieved by one or more of the following strategies:

a. RNA H independent ODNs: Ribonuclease (RNA H), which is independent of oligodeoxiribonucleotides (ODN). Oligodeoxiribonucleotides are sometimes called oligonucleotides, which is a short sequence of nucleotides (RNA or DNA) typically with twenty or fewer base pairs.
b. RNA H dependent ODNs
c. Ribozymes and DNA enzymes
d. siRNA, which is a small interfering RNA (siRNA), sometimes known as short interfering RNA or silencing RNA, is a class of double-stranded RNA molecules, 20-25 bp in length, that play a variety of roles in biology. Most notably, siRNA is involved in the RNA interference (RNAi) pathway, where it interferes with the expression of a specific gene.
e. Methylation of DNA
f. Drugs

Proteins are made out of amino acids through the process of transcription of DNA to RNA and the translation fro RNA to protein, as discussed before. The DNA sequence in genes is copied into a messenger RNA (mRNA). Ribosomes then translate the information in the mRNA and use it to produce proteins from amino acid which are carried by tRNA (transfer RNA) which enters the ribosome and meets the mRNA, Figure (68).

Figure (68): Gene silencing

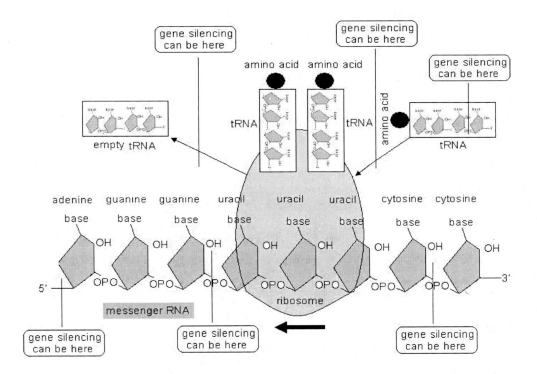

Ribosome is like credit card, when you swipe it, it reads the memory. It reads both tRNA and mRNA to produce protein. If there is an interruption (silencing) to the pathway of the tRNA or mRNA, the protein will ne damaged or another type of protein is produced. Protein output can be modulated to fight the disease such as AIDs or cancers. Gene silencing can also happen at the prime (5' or 3')of the DNA, since the equation of protein's output is :

$$\text{DNA} \xrightarrow{\text{transcription}} \text{RNA} \xrightarrow{\text{translation}} \text{Protein}$$

3-28-3 Gene Silencing and Aging

Gene silencing in aging is a theory which says that aging relates to silencing of genes involved in the control of cell cycle, apoptosis, detoxification, and cholesterol metabolism. Many studies consider the most important of these the silencing is tumor suppressors.

The fact that there are substantial gene expression changes due to silencing related to aging has been confirmed in studies by different groups, conducted in yeast, worms, flies and mice. An aging mechanism is proposed based on this gene-silencing phenomenon whereby accumulation over time of methylation, oxidation, and acetylation of histones contributes to cellular senescence. Over time, this process of methylation, oxidation and acetylation of histones may lead to widespread gene silencing in diverse dividing and nondividing cell types contributing to aging of the organism.

3-28-4 Acetylation of histones

Histones are the proteins closely associated with DNA molecules and they are the chief protein components of chrpomatin as explained earlier. Acetylation of histone protein removes positive charges, thereby reducing the affinity between histones and DNA. This makes RNA polymerase and transcription factors easier to access the promoter region. Therefore, in most cases, histone acetylation enhances transcription while histone deacetylation represses transcription. Acetylation or ethanoylation can be achieved by the addition of acetyl group $COCH_3$ to the histone as ethanol has two atoms of carbons, see chapter 2.

Acetylation brings in a negative charge, acting to neutralize the positive charge on the histones and decreases the interaction of the N termini of histones with the negatively charged phosphate groups of DNA. As a consequence, the condensed chromatin is transformed into a more relaxed structure which is associated with greater levels of gene transcription. This relaxation can be reversed by HDAC (histone deacetylate) activity. Relaxed, transcriptionally active DNA is referred to as euchromatin. More condensed (tightly packed) DNA is referred to as heterochromatin. Figure (69) shows acetylation and deacetylation of histone lysine.

Figure (69): Condensed and relaxed DNA due to acetylation and deacetylation

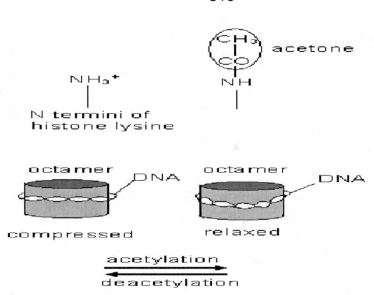

Greater level of gene transcription can elongate the life span and can be regulated to fight diseases.

Histone acetylation is catalyzed by histone acetyltransferases (HATs) and histone deacetylation is catalyzed by histone deacetylases (HDs). Several different forms of HATs and HDs have been identified. Among them, CBP/p300 is probably the most available, and can interact with numerous transcription regulators. The unusual properties of p300/CBP is one of the most important enzymes in the HAT family, and can be targeted for developing new anti-cancer drugs, says Ronen Marmorstein, Ph.D., a professor in the Gene Expression and Regulation Program at Westar.

3-29 Facts about aging

3-29-1 Changes that Typically Occur with Age, Table (6).

Table (6): Typical changes occurred with aging

Sooner or later, many people develop one or more of the degenerative diseases of aging, such as:
- Alzheimer's disease
- Stroke
- Heart attack
- Cancer
- Osteoporosis

- Diabetes mellitus (the body may produce insufficient amounts of insulin to process blood glucose, or the body's tissues may not use insulin properly)
- Parkinson's disease (Parkinson's disease is a disorder that affects nerve cells, or neurons, in a part of the brain that controls muscle movement. In Parkinson's, neurons that make a chemical called dopamine die or do not work properly. Dopamine normally sends signals that help coordinate your movements. No one knows what damages these cells).
- Arthritis
- Cataracts
- Glaucoma (Glaucoma is a group of diseases that affect the optic nerve and involves a loss of retinal gangalion cells in a characteristic pattern. It is a type of opyic neuropathy).
- Hearing loss
- These conditions cause pain and suffering to patients and their families - the economic and social costs are staggering

3-30 Arguments about aging

a. Our body has been selected in order to survive until the age of reproduction and the average life expectancy at birth did not exceed 40-45 years until a couple of centuries ago even in the most developed countries, and our genes and gene variants have been selected to survive in such a harsh environment until the age of reproduction.
b. Infectious diseases exert a constant evolutionary pressure on the genetic makeup of our innate immune system. Therefore, our immune system has been molded by evolution and it is quite efficient in coping with acute infections in young people. This immune system is not efficient in responding to diseases at aging period.
c. Our immune system (both innate immunity and adaptive immunity) appear to be hyper-stimulated in old people owing to decades of evolutionary unpredicted antigenic stimulation, mostly due to subchronic bacterial and viral infections. Our immune system has evolved to control pathogens, so pro-inflammatory responses are likely to be evolutionarily programmed to resist fatal infections with pathogens aggressively. Thus, inflammatory genotypes are an important and necessary part of the normal host responses to pathogens in early life, but the overproduction of inflammatory molecules might also cause immune-related inflammatory diseases and eventually death later.
d. All these phenomena have a strong and variable genetic component as shown by studies in centenarians which collectively showed that the frequency and duration of several variants (polymorphisms) of important molecules involved in immune response and inflammation are different in long lived people in comparison to young subjects.

e. Within this scenario the age-related changes in physical body composition and cognitive power, and the cross talk between macrophages and adipocytes (macrophages are believed to alter insulin sensitivity in adipocytes), and alterations of signaling pathways such as the IGF1/insulin pathway, individual's genetic make-up and the function of critical organelles such as the mitochondria are also emerging as critical target for the development of frailty and major age-related diseases.

3-30-1 Social argument about gerontology

a. Social gerontology emerged as a field of study which attempted to respond to the social policy implications of demographic change (Vincent 1996). Such disciplines were shaped by significant external forces such as economy, religion and culture. First, by state intervention to achieve specific outcomes in health and social policy; secondly, by a political and economic environment which viewed an aging population as creating a 'social problem' for society (Jones, 1993); thirdly, by development and evolution of education and technology.

b. Two theories have been established:

b.1 The disengagement theory was suggested by the structural-functionalist school via the work of the theorists Phillipson, 1998, and Cumming and Henry, 1998. They asked that older people should disengage from work roles and prepare for the ultimate disengagement: death (Powell 2000), and should also withdraw themselves from work roles and social relationships. The theorists viewed that as inevitable and natural process. This theory argued that it was beneficial for both the aging individual and society that such disengagement takes place in order to minimize the social disruption caused at an ageing person's eventual death (Neurgarten 1998). A number of critiques exist: firstly, this theory condones indifference towards 'old age' and social problems. Secondly, the disengagement theory underestimates the continuity of cultural and economic structures old people are involved in. Also, Kastenbaum (1993) claims disengagement theory represented a threat to the promotion of a positive and involved special lifestyle for ageing persons across the life course.

b.2 Activity theory is a counterpoint to disengagement theory, since it claims a successful 'old age' can be achieved by maintaining roles and relationships. Ageing can be a lively and creative experience. Any loss of roles, activities or relationships within old age, should be replaced by new roles or activities to ensure happiness and continuation of smoothness of life, value consensus and well-being. Thus, "activity" was seen as an ethical and academic response to the disengagement theory which saw retirement as joyous and mobile. Argument between the two theories created a feeling of redundancy in the social structure by demanding you, as an aged person, will either 'disengage' or will be 'active'. This can be argued to be a form of 'academic imperialism'.

c. Political economy

Older people are viewed as a burden on western economies, with demographic change...They are seen as creating intolerable pressures on public expenditure. This is due to dominance of structural functionalism in the form of disengagement theory, the biomedical expenses and world economic crises of the 1970s. The starting point for policy-makers should be the labour market and the social relationship between age and the labour market. Elderly people have incomes equal to or below the poverty line compared with non-elderly people due to their less participation in the labour market. In Britain, nearly two-thirds of the elderly, comprising 5.1 million people, live in or on the margins of poverty, compared with one-fifth of the non-elderly. In the USA about one in every five elderly people have incomes below the federally established minimum. In Japan, nine-tenths of elderly people have incomes in the lower half of the income distribution. While the major problem of poverty in old age has been ignored in recent official research and policy statements, low incomes have been accepted as an inevitable consequence of old age and therefore something to which elderly people must 'adjust'. This acceptance is legitimated through theories of need based partly on age and narrowly functionalist theories of ageing.

Approaches to age and ageing based on the implicit assumption that the elderly can be treated as a distinct social group, in isolation from the rest of the social structure, have provided a totally inadequate basis for an explanation of the persistence of poverty in old age, and continue to obstruct the formation and application of social policies aimed at solving this problem. After lengthy discussions to solve the problem, two new strategies have been put forth:

1- to orient priorities from "in-home" care for the elderly by family members to care taking by the municipal government institutions,
2- with regard to incapable aged people who are living alone, or with a partner (usually a spouse) to provide them also with improved home services such as providing them with a support of salary (old age pension).

In both cases, welfare services were adopted to provide welfare programs for aged people at their homes or at municipal governmental institutions. There is a third orientation sponsored by private-sector welfare services as the private sector should be encouraged to perform a greater role of providing such services.

In developing countries, including the agrarians, old people are not considered as social agents. They are turned out to be symbolic objects that are respected and mostly cared by their children. Modernization of societies transforms social archetype and cultural standard from long–established values and respects of old people to less and weaker interest.

d. Social democrats for aged people

The post-1945 welfare state settlement is no longer adequate for today's world. The scale of the future challenges is self-evident: the intensification of international competition, an ageing population, changing gender roles in households and labour markets, new technologies, rising social tensions and deeper inequalities. If the centre-left merely defends existing welfare state entitlements and distributive arrangements, it will be incapable of responding adequately.
Instead, social democrats have to begin by shaping a new political strategy as distinct from the policy agenda.[1] This involves four strategic shifts:

1.   Social democrats need a progressive narrative that connects economic and social policy. At the heart of the social democratic argument is the idea that a strong welfare state complements, and is a prerequisite for, an enterprising economy. That requires active labour market policy, childcare, and work-life balance as the route to economic success. As Göran Persson, the Swedish Prime Minister puts it, "the welfare state contributes to people's freedom and enterprise. It is people that are secure, that dare to try new things".

2.   Social democrats have to focus on building progressive institutions for the future, as they have done for the past half-century: an inclusive pension system, universal childcare, strong parental leave provision. In the UK, the Labour Government is building a national infrastructure of childcare centres. These not only provide support for parents and children, but also embed social capital. They hard-wire social democratic values into the fabric of society, and are unlikely to be dismantled or undone by future governments.

3.   In a globalizing world, social democrats have to fashion policy instruments and co-ordinate action across the traditional boundaries of the nation-state. The aspiration for social democratic 'Keynesianism in one country' has been proved illusory. The EU and the international institutions more generically are increasingly pivotal to realizing our ambitions for social justice in the 21st century.

4.   Finally, social democrats must entrench welfare states that cater to the needs of the majority, not simply the minority who are poor. In areas such as childcare and pensions, it is necessary to give the majority a stake in progressive institutions.

e. New challenges of the welfare state

Enormous and continuous actuarial and structural challenges are required in order to cope with the coming crises:

1.   On average, 16% of the populations (on the two sides of the Atlantic) are over 60 years, people are living longer, but the main problem is the low birth rate.
2.   Obese people in the state are 25 per cent today compared to those in the EU. Obesity leads to diabetes and other diseases which placing heavy burden on healthcare systems and consequently on the economic situation. Generally, chronic diseases such as smoking, heart disease, diabetes, and cancer – which make up about 75 per cent of all diseases are caused by lifestyle rather than infection.
3.   Some studies recently showed that there are new phenomena of poverty and deprivation in all the industrialized countries due to immigrants, legal or otherwise, vulnerable to exploitation, fractured families and weak family substitutes, the mentally ill, and victims of violence, especially women who suffer domestic violence. In some western countries, women violence is the leading cause of death – higher than car accidents or cancer.
4.   Rates of divorce are higher, and rates of marriage lower, than in the past. Families, in most EU countries, are much more mobile and lack the extended kin relations that were once a source of social support. Western societies are also witnessing a rise in the proportion of 'non-conventional families': women having children alone, and same-sex couples. In 1980 in the UK, 12 per cent of all births were outside marriage; in 2004, this had increased to 42 per cent. See www.youngfoundation.org.uk.
5.   There has been a sharp decline in the growth potential of the western world. For example, in the EU the decline dropped from 2.4 per cent in the early 1990s to less than one per cent in 2003.

This will affect the productivity and efficiency since there will be a drift in the labor market towards old people.

In spite of above challenges and problems, there is a promising hope that there has been a vast interest in postmodern perspectives of age and aging identity underpinned by discourses of "better lifestyles" and increased leisure opportunities for older people due to healthier lifestyles and increased use of bio-technologies to facilitate the longevity of human experiences

3-31 Author's Exclusive Opinion about Causes of Aging and Elongation of Life Span

3-31-1 Ultraviolet rays and aging

The Sun emits ultraviolet radiation in the UVA (400-320nm), UVB (320-280nm), UVC (100-28nm) and extreme EUV (10-121nm) bands. The Earth's ozone layer blocks 98.7% of this UV radiation from penetrating through the atmosphere. 98.7% of the ultraviolet radiation that reaches the Earth's surface is UVA.

The human cell diameter is about 80nm. Therefore, the EUV which is the fastest can penetrate the human skin cell and kill it and also damage the DNA of the cell.

UV exposure decreases rapidly at increasing depths in the water column. In other words, water and the impurities in it strongly absorb and scatter incoming UV radiation. Some substances that are dissolved in water, such as organic carbon from nearby land, will also absorb UV radiation and enhance protection of microorganisms, plants, and animals from UV. Different masses of water at different locations contain different amounts of such dissolved substances and other particles, making the damage of UV minimal.

Therefore, lobsters, alligators, sharks, and turtles show few signs of aging.

Conclusion: Exposure to ultraviolet rays can speed aging. Therefore, for longer life span avoid UV rays.

3-31-2 Sugar and protein

The chemical formula of protein ranges from $C_2H_5NO_2$ (glycine) to $C_{5978}H_{9285}N_{1835}O_{1941}S_{31}$ (Enaptin). There are several hundreds of thousands of different proteins, a protein is any chemical made of only amino acids strung together in a long chain. The noticeable thing in the chemical structure of proteins is that all of them have much less oxygen than carbon except for the glycine. In sugar, Carbon and oxygen atoms are more or less equal. So if you combine sugar and proteins chemically, say using heat, there is a trend for the oxygen to oxidize the protein. The Millard reaction is a chemical reaction between an amino acid and a sugar, usually requiring heat. It is a step in the formation of advanced glycation endproducts. The Maillard also showed a correlation in numerous different diseases in the human body, particularly degenerative eye diseases. These diseases are generally due to the accumulation of advanced glycation end products, or AGEs, on DNA, RNA, proteins, and lipids.

Conclusion: Avoid eating high proportion of sugar to protein. Protein should be much more than reduced sugar (glucose, fructose, glyceraldehyde, lactose, arabinose and maltose). Seek more protein and less sugar. The proportion should be in the range of 20 grams of protein to 1 gram of sugar.

3-31-3 pH level and aging

Intracellular acidosis (low pH) can activate protein p38 MAPK. The University Of North Carolina School Of Medicine at Chapel Hill reported that the p38MAPK protein, already known for its role in inflammation, also promotes aging when it activates another protein p16, which has long been linked to aging. Acidosis pH means more hydrogen added to the metabolism. Hydrogen (proton) is characterized by a positive charge which can be easily covalently attached to the negative oxygen in the DNA, and consequently either damage the DNA or delay the process of transcription and translation. Hydrogen can also combine with histones (which terminate with negative NO^2 or negative NO^3) in the chromatin and damage the chromosome and the emerging DNA.

The elderly are more prone to develop acid-base disturbances than the young. With age, the kidney undergoes structural and functional changes that limit the adaptive mechanisms responsible for maintaining acidbase homeostasis in response to dietary and environmental changes.

People with Type I and type II diabetes lack enough insulin, a hormone the body uses to process sugar (glucose) for energy. When glucose is not available, body fat is broken down instead. The byproducts of fat metabolism are ketones and acid. When fat is broken down, ketones and acid build up in the blood. A condition called ketoacidosis develops when the blood has more acid than normal.

Conclusion: Eat fresh fruits and vegetables, whole grains, legumes, and nuts and seeds because they are consistently on the top of the alkaline list (high pH). While processed foods, red meat, white flours and refined sugars, part of the typical diet, are definitely acidic (low pH).

Patients with diabetes should keep the sugar level in their blood as normal as possible. Otherwise, pH level will go below 7 due to the accumulation of ketoacidosis.

3-31-4 Cholesterol and albumin

Men with low cholesterol and low albumin were more likely to experience a decline in physical performance. Low serum levels of albumin, total cholesterol and iron have been shown to predict morbidity and mortality, and may reflect frailty in older individuals. Ovalbumin is the main protein of egg white, lactalbumin occurs in milk, and plasma or serum albumin is one of the major blood proteins.

Conclusion: Eggs supply all essential albumin and amino acids for humans, and provide several vitamins and minerals, including vitamin A, riboflavin, folic acid, vitamin B6, vitamin B12, choline, iron, calcium, phosphorus and potassium. All of the egg's vitamin A, D and E is in the egg yolk.

3-31-5 Cysteine

Food sources of cysteine include poultry, yogurt, egg yolks, red peppers, garlic, onion, broccoli, brussel sprout, oat and wheat germs. Cysteine is a powerful antioxidant on its own. When taken orally, cysteine is believed to help the body make another important antioxidant enzyme glutathione. It has shown promise for a number of conditions such as bronchitis, angina, acute respiratory distress syndrome, kidney damage, colon cancer prevention and aging. It has also been claimed that L-cysteine has anti-inflammatory properties, that it can protect against various toxins, and that it might be helpful in osteoarthritis and rheumatoid arthritis.

Conclusion: Increase cysteine in your diet.

3-31-6 Coenzyme A (CoA)

The pyruvate formed during glycolysis is delivered into the mitochondria, where it is further broken down into another intermediate called acetyl CoA. This process releases some electrons which are harvested for use in electron transport by charging NAD^+ into NADH. Acetyl CoA is then used again in Krebs cycle (or citric acid cycle) to liberate many times more electrons for formation of more and more NADH. This process of transportation of more electrons generates proton reservoir that drives the generation of nearly all the cell energy. In the reverse process which is the conversion of NADH back to NAD^+, there are less electrons and more hydrogen. This redox process maintains the body in a balanced state. This is what happened in muscle cells during aerobic exercise. With intense anaerobic exercise, the cell is working hard and uses up all of the available oxygen, and can't keep oxidative phosphorylation going. So, with aerobic exercise there is a balanced state between the ATP and the glycolysis. With anaerobic exercise, the phosphorylation is not completed and thus the pyrovate converts into lactic acid, causing the pH to reduce, the synthesis of ATP to slow down, and the P/O ratio (phosphate per oxygen ratio) to reduce.

Conclusion: Performance of aerobic exercises can burn the sugar in a balanced state, i.e. excess sugar is burnt and not stored as fat. Anaerobic can store the excess sugar as fat.

3-31-7 Sex, aging and longevity

Despite a large number of studies, available data do not allow at present to reach definitive and clear conclusions on role of sex on longevity. However, recent studies showed that excessive sexual activity and over-ejaculation lead to overproduction of androgen hormones, causing adrenal and sex organ fatigue, and excess release of dopamine to maintain prolonged sexual arousal. Since dopamine is the precursor to the stress hormone epinephrine (adrenaline), excess dopamine results in uncontrollable/subconscious movements (like picking, tapping, repetitive moments, jerking, twitching). Remember that the heart is a muscle, too, and too much dopamine will result in increased pulse and blood pressure. The adrenal glands overproduce epinephrine and putting the body in a prolonged state of fight-or-flight stress. At the same time, norepinephrine is synthesized from dopamine and released from the adrenal medulla into the blood as a hormone, along with the stress hormone cortisol. Epinephrine, norepinephrine and cortisol fuel the fight-or-flight response, directly increasing heart rate, triggering the release of glucose from energy stores, and increasing blood flow to skeletal muscle. All of this has a severely taxing effect on the body. With repetitive sexual intercourse, the brain may shut down dopamine receptors and may cause the brain to lose its ability to produce high levels of dopamine.

Conclusion: Too much sex would speed aging, increase blood pressure, and heart disease.

3-31-8 Alcohol and aging

Aging and alcoholism produce similar deficits in intellectual (i.e., cognitive) behavioral, and physical functioning. Alcoholism may accelerate normal aging or cause premature aging of the brain. Using magnetic resonance imaging techniques, it was found that more brain tissue loss in subjects with alcoholism than in those without alcoholism, even after their ages had been taken into account. In addition, older subjects with alcoholism exhibited more brain tissue loss than younger subjects with alcoholism, often despite similar total lifetime alcohol consumption. These results suggest that aging may render a person more susceptible to alcohol's effects, Oscar-Berman, M.; Shagrin, B.; Evert, D.L.; et al. Impairments of brain and behavior: The neurological effects of alcohol. Alcohol Health Res World 21(1):65-75, 1997.

Alcohol intoxication activates the HPA (hypothalamic-pituitary adrenal) axis and results in elevated glucocorticoid levels. The glucocorticoid is a stress hormone in the body. Chronic excessive glucocorticoid secretion can have adverse health effects, such as Cushing's syndrome, and can result in premature and /or exaggerated aging.

For people on certain medications, it may be dangerous to drink any alcohol. Older drinkers are likely to be on medications. Keeping balance while walking or standing becomes more difficult with age and alcohol makes the problem worse. Falls and other injuries are more common with alcohol use.

When you drink an alcoholic beverage around 2 to 8 percent is lost through urine, sweat, or the breath. The other 92 to 98 percent is metabolized by your body. All ethyl alcohol which is broken down in the human body is first converted to acetaldehyde, Figure (70), and then this acetaldehyde is converted into acetic acid radicals--also known as acetyl radicals. Acetaldehyde is a poison which is a close relative of formaldehyde. Too much acetaldehyde may cause deacetylation to histons and result in reduction of transcription of the DNA, which in turns results in reduction of RNA translation, i.e. less protein.

Figure (70): Conversion of alcohol to acetaldehyde

ethyl alchol ... enzyme → acetyldehyde ... + 2H

C2H6O   C2H4O

Conclusion: Stop drinking and don't drink alcohol if you are over 65.

3-31-9 Electromagnetic fields and aging

Electromagnetic radiation can be classified into ionizing radiation and non-ionizing radiation, based on whether it is capable of ionizing atoms and breaking chemical bonds. Both electromagnetic types could be associated with three major potential hazards: electrical, chemical or biological.

- The IEEE (USA), IEE (UK), Cigre (France), VDE (Germany), JIS (Japan) and many other international associations have established safety limits for exposure to various frequencies and voltages of electromagnetic fields. Here are some limitations:
- For residential exposure – 1000 milli gauss (unit of measurement of magnetic flux density)
- For occupational exposure – 5000 milli gauss
- Power line should not exceed 1000 milli gauss for 24 hours,
- Magnetic fields in schools, daycares, and play grounds should not exceed 2-3 milli gauss (Sweden).

In Europe, loads of non-linear type (inductive or capacitive, but not resistive) must have filters built into it at the factory. As a result, American computer manufacturers must put an inexpensive RF (radio frequency) filters in the computers they export to Europe.

Conclusion: There are different and inexpensive meters in the market which can measure the gauss level. Buy one and measure the gauss level radiated from your microwave, lighting dimmers, discharge lighting fixtures, motors, and sources of electrical noises in your home.

3-31-10 Sleep and aging

Sleep allows our body to rest and to restore its energy levels. Without enough restful sleep, not only can we become grumpy and irritable, but also inattentive and more prone to accidents.

As we age, our bodies secrete less of two important sleep hormones: melatonin and growth hormone. Melatonin is important because changes in the level of this hormone control the sleep pattern. With less melatonin, many older adults feel sleepy in the early evening and wake up in the early morning. They also may have more trouble falling asleep. The growth hormone is what makes children sleep so deeply. As we age, our body secretes less of this hormone and deep sleep becomes more difficult. Menopause causes a great deal of hormonal changes in women, sometimes resulting in night sweats and other symptoms that interfere with sleep.

No matter what our age, sleeping well is essential to our physical health and emotional well-being. As we age, a good night's sleep is especially important because it improves concentration and memory formation, allows your body to repair any cell damage that occurred during the day, and refreshes your immune system which helps to prevent disease.

Conclusion: Try making changes in your sleep and life style habits if you don't sleep enough. If your age is 60 and over, you need between 7 to 9 hours of sleep each night. Don't take medication for sleep. Try to limit caffeine late in the day, avoid alcohol before bedtime, satisfy your hunger prior to bed, avoid big meals or spicy foods just before bedtime, minimize liquid intake before sleep, and combine sex and sleep as sex and physical intimacy, such as hugging and massage, can lead to restful sleep.

3-31-11 Superoxide dismutase, catalase, and glutathione peroxidase in red blood cells

SOD, CAT, and glutathione can be increased by increasing the number of blood cells.

- Eating more iron rich foods or an iron supplement can promote red blood cell production. On the flip side, not having enough iron in your diet can lead to a red blood cell deficiency.
- Exercise regularly as exercising on a regular basis increases red blood cell count.
- Consume the proper levels of Vitamin B-12 and folate. These compounds are important in the production of red blood cells.

Conclusion: Red blood cells are associated with Superoxide Dismutase, Catalase, and Glutathione Peroxidase which are decreased with aging. Superoxide Dismutase, Catalase, and Glutathione Peroxidase are vey effective scavengers of radicals.

3-31-12 Radiation

Beta ray, X-ray, Gamma ray, and neutrons ray as all of them penetrate the human body, and could damage the cells and genetic components such as DNA and RNA, Figure (71).

Beta rays are produced following spontaneous decay of certain radioactive materials, such as tritium (an isotope of hydrogen), carbon-14, phosphorus-32, and strontium-90. Depending on its energy (i.e., speed), a beta ray can traverse different distances in water--less than 1 mm for tritium to nearly 1 cm for phosphorus-32. As with alpha rays, the major concern for health effects is after their ingestion (i.e., internal exposure).

Gamma rays are produced following spontaneous decay of radioactive materials, such as cobalt-60 and cesium-137. A cobalt-60 gamma ray can penetrate deeply into the human body, so it has been widely used for cancer radiotherapy.

X rays have the same characteristics as gamma rays, although they are produced differently. When high-speed electrons hit metals, electrons are stopped and release energy in the form of an electromagnetic wave. This was first observed by Wilhelm Roentgen in 1895, who considered it a mysterious ray, and thus called it an X ray. X rays consist of a mixture of different wavelengths, whereas gamma-ray energy has a fixed value (or two) characteristic to the radioactive material.

Neutron particles are released following nuclear fission (splitting of an atomic nucleus producing large amounts of energy) of uranium or plutonium. In fact, it is neutrons that trigger the nuclear chain reaction to explode an atomic bomb. The human body contains a large amount of hydrogen (a constituent of water molecules that occupy 70% of the human body), and when neutrons hit the nucleus of hydrogen, i.e., a proton that is positively charged, the proton causes ionizations in the body, leading to various types of damage. At equivalent absorbed doses, neutrons can cause more severe damage to the body than gamma rays. (Neutrons hardly damage cells because they do not carry any electrical charge.)

Conclusion: Avoid such rays which could damage the cells and genetic components.

Figure (71): Penetration of alpha, beta, X-rays, gamma-rays, and neutrons

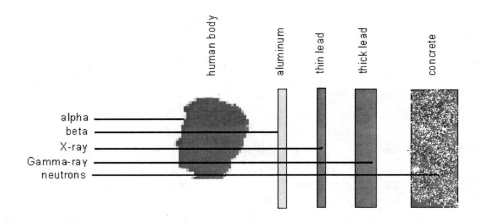

3-31-13 Electrical pulses, impulses and voltage surges

Pulses, impulses, and voltage surges are brief sudden changes in a normally constant quantity such as current or voltage or any of a series of intermittent occurrences characterized by a brief sudden change in a quantity, Figure (72). Studies show that pulses of 5kV and duration 5.8 ms (millisecond) have direct proportionality with cancer. It means that cancerous tumors are increased if the value of voltage and duration are increased above the 5kV and 5.8 ms. In other words, if the pulse energy is increased, the tumor will increase.

Figure (72): Types of disturbances that affect cancerous tumors.

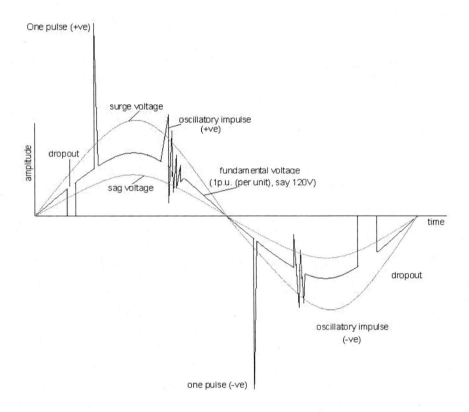

Pulses and impulses can be removed from the electrical system by using proper transducers to clamp the increased voltage above the fundamental value. Transducers such as varistors (metal oxide varistor; MOV), or capacitors, or combination of varistors and capacitors are available at electronic shops. Some surge protectors are equipped with such components, Figure (73).

Figure (73): Clamping of pulses by MOV (metal oxide varistor)

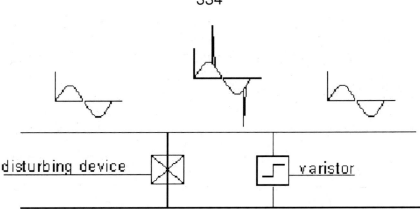

<u>Precaution:</u>
For safety reasons, don't use, connect, or weld such components unless you are a licensed electrician/technician.

Electrolytes and electrical pulses. It has been proposed that electrical pulses could affect the level of electrolyte, and the result is disturbed electrolytes. For example, elevated levels of K^+ ions and Ca^{2+} ions increase adenosine 3', 5'-monophosphate (cyclic AMP) levels in brain tissue *via* release and subsequent action of adenosine. Adenosine will be separated from cyclic AMP due to the combination of the O^- and the ions released from pulses. Cyclic adenosine monophosphate is a second messenger that is important in many biological metabolisms. It is a part of the ATP and could have an effect on the sugar and glycogen in the liver. Cyclic AMP is also used to regulate Ca2+ passage into the brain. Electrolytes (Na^+, K^+, Ca^{2+}, Mg^{2+}, Cl^-, PO_4^{3-}, and HCO_3^-) can be elevated (hyper) or depleted (hypo) du to the effect of electrical disturbances such as pulses and impulses.

Electrolytes are important for the metabolisms and for the functions of neurotransmitters because they maintain voltages across cell membranes in order to carry proper electrical signals to nerves, muscle contractions, and to maintain balance between body's senses. Kidney is the most important organ to keep balanced electrolytes.

Dr. Payne and her co-investigators from Duke and the University of North Carolina examined magnetic resonance imaging (MRI) scans from 232 men and women (79 men, 153 women) between the ages of 60 and 86 (average age 71). All the subjects had at least some brain lesions of varying sizes, including the extremely miniscule ones often seen in even healthy older persons, but those who reported consuming more calcium and vitamin D were markedly more likely to have higher total volume of brain lesions as measured across numerous MRI scans, http://www.medicalnewstoday.com/articles/69610.php.

Pulses can have an effect on the shape of hyper and hypo electrolytes. For example, hyperkalemia (extra potassium) has a "tent shape" in normal conditions. When the body exposed to high electrical pulses, the hyperkalemia changes its pattern to T or U shape which are the characteristics of hypokalemia. When untreated, hypokalemia may lead to arrhythmia which is related to heart disease.

Pulsed also affect the ATP and ADP of the sugar cycle, and accordingly the quantity of fat and sugar which consequently affect the amount of insulin and glucagon released by the pancreas that could lead to pancreatic stress. Hypo and hyperkalemia can also affect the QT intervals of the heart pulses which could lead to long QT syndrome.

Conclusion: Don't expose yourself to high pulses which could damage the kidneys and liver due to imbalance in the electrolytes.

3-31-14 Depression

Depression often occurs during the aging process but doesn't always have to be. Many times people will believe that they are depressed because they are aging but most often depression is just a symptom of some other physical or emotional problem which can also happen with adults.

Unfortunately, the aging process is not always so pleasant. Late-life events such as chronic and debilitating medical disorders, loss of friends and loved ones, and the weakining to take part in once-cherished activities can take a heavy toll on an aging person's emotional well-being.

An older adult may also sense a loss of control over his or her life due to failing eyesight, hearing loss, and other physical changes, as well as external pressures such as limited financial resources. For many people aging means losing their independence and having to rely on the help of others more and more. These and other issues often give rise to negative emotions such as sadness, anxiety, loneliness, and lowered self-esteem, which in turn lead to social withdrawal and apathy. The important thing to remember is that feeling some sadness about aging is perfectly natural. It's how you handle this sadness that is important.

Aged people show a variety of alterations in hypothalamic-pituitary-adrenocortical (HPA) system regulation which is reflected by increased pituitary-adrenocortical hormone secretion. Major alterations of the hypothalamic–pituitary–adrenocortical (HPA) system that can be reversed by successful antidepressant therapy are often seen in depressed patients.

Conclusion: There are many steps you can take. Try to prepare for major changes in life, such as retirement or moving from your home of many years. One way to do this is to try and keep friendships over the years. You can also develop a hobby. Hobbies may help keep your mind and body active. Stay in touch with family. Let them help you when you feel very sad. Regular exercise may also help prevent depression or lift your mood if you are somewhat depressed. Older people who are depressed can gain mental as well as physical benefits from mild forms of exercise like walking outdoors or in shopping malls. Gardening, dancing, and swimming are other good forms of exercise. If depression persists, see a health provider.

3-31-15 Hormones treatment

For more than a decade, the National Institute on Aging (NIA), a component of the Federal Government's National Institutes of Health, has supported and conducted studies of replenishing hormones to find out if they may help reduce frailty and improve function in older people. These studies have focused on hormones known to decline as we grow older:

- Growth hormone
- Menopausal hormones
- Testerone
- Melatonine
- Dehydroepiandrosterone (a hormone that is abundant at infancy and young adulthood. Some believe that supplements of DHEA may increase youthful vigor, but the effects of such supplements are not yet understood)

Conclusion: Don't take any of the above hormones as the NIA has not yet approved them. Recommendations to use supplemental hormones and hormone-like molecules to influence the aging process and health problems associated with aging should be viewed with skepticism.